Columbia Food

Columbia Food

A HISTORY OF CUISINE IN THE FAMOUSLY HOT CITY

LAURA ABOYAN

AMERICAN PALATE

Published by American Palate
A Division of The History Press
Charleston, SC 29403
www.historypress.net

Copyright © 2013 by Laura Aboyan
All rights reserved

First published 2013

ISBN 978.1.60949.819.1

Library of Congress CIP data applied for.

This book is dedicated to my Nonie. Even though food was not her forte—she once ordered a swordfish steak and was surprised when she got fish instead of red meat—I like to think she would have appreciated this book. I miss you Nonie. I do hope I have made you proud.

CONTENTS

CONTENTS

ACKNOWLEDGEMENTS

I guess I should start by thanking the Columbia Visitors' Bureau for running a contest for restaurant week in 2011. If I had not been lucky enough to win the contest, I never would have started a food blog. And if I had never started "The Hungry Lady," I would never have been approached about writing a book about food. CVB, you are the snowball that started the avalanche. Thanks so much!

Of course, major thanks are due to all the people who made this book possible and were kind enough to share their experiences and their passions for local food. From chefs to local business owners to my fellow bloggers and local residents, your thoughts were much appreciated.

To all of the readers of "The Hungry Lady" and the Twitter and Facebook followers, thank you so much for all of your comments, feedback and genuine excitement about this project, as well as really just for being sweet enough to read my blog. Your insights and contributions to this project were invaluable.

My undying gratitude goes to Clarissa Johnson for taking such wonderful photographs and for putting up with my disorganization. Similarly to my commissioning editor, Chad Rhoad, thank you so much for putting up with my shenanigans. I cannot even imagine the number of headaches I caused you during this entire process. Thanks for digging deep, finding a little faith and sticking with me through it all.

Many thanks to my wonderful group of friends. They are kind enough to cook for me or join me out on the town for various meals or a much needed happy hour. I know I am a terrible dining companion, what with the picture taking and the constant analysis of the food. I am not quite sure why you put up with me, but I sincerely appreciate it.

Finally, I am extraordinarily lucky to have a family who has always told me that I could do anything I put my mind to, who supported me while I was trying not to fall on my face and who were there to pick me up when I did, in fact, fall flat. I would never have made it through this process without their support, encouragement and love—well, that and their constant barrage of questions: How is the book coming? When is it being released? You are going to thank me in your acknowledgements, right? That last one was totally asked by my mom. She was only kind of kidding when she asked it. Right. Like I would actually forget to thank my mom. She has always been the first person to push me out of my comfort zone, to encourage me to try for the seemingly impossible and to celebrate my accomplishments. She let me rage at her for hours about my frustrations, virtually hugged me and wiped away tears and shared a happy hour beer over the phone when I was feeling discouraged. There is absolutely no way that this book would have come to fruition if it had not been for my utterly amazing mother. So thanks, Mom. I love you. And I am sorry that this thank-you has made you cry.

WELCOME TO THE JUNGLE

If you survey foodies today, you will hear an awful lot of buzzwords. Locavore. Organic. Locally grown. Grass-fed. In the last twenty years or so, supporting local food has become a movement, and one that I fully support, but it is not a lifestyle into which I had ever put a conscious effort. I grew up in a rather unpretentious suburb in Pennsylvania that turned quickly from single-family homes to rambling fields and farmland. There were always fresh vegetables and fruits to be had—roadside stands were prevalent on even the busiest of main roads. The farmers' market was around the corner, and it seemed that everyone had their own garden (except my family—we are definitely not known for our green thumbs). Eating seasonally and locally wasn't a movement to us; it was just the way everyone lived.

Fast-forward a few years, and I was living far away from that "living off the land" mentality. Sure, there were still options for buying food with all of the aforementioned buzzwords attached, but I had just finished college, and my options were to either pay my rent or to spend my meager paycheck on locally grown and/or organic food. Homelessness did not hold much of an appeal to me, so my landlord stayed happy while I bargain-shopped for food that was not terrible for me and daydreamed about the time when I could go to some of those places mentioned on the countless episodes of *Top Chef* that I watched that specialized in seasonal and locally grown food.

Slide a few more years down the "I am rapidly becoming an adult but would rather run off to Neverland" timeline, and I found myself in possession of an absurd number of gift cards to some of the best restaurants in Columbia. What a pleasant surprise to find that in a fertile bed of chain restaurants and fast-food joints, there were actually places that devoted their menus to using the best seasonal ingredients they could find from some of the best local vendors in the area. Suddenly, all of the places I had heard mentioned in the hushed and reverent tones usually reserved for trips to museums or libraries were at my fingertips.

I dove headfirst into the world of local dining, trying foods I never would have been exposed to otherwise. Sausage and quail and duck from local farms. Fish caught off the coasts of South Carolina. Sauces, garnishes and sides made from vegetables that had been harvested the day before. Desserts that changed based on the season. I discovered a love for beets, a food that I had not eaten since I was an infant. Always a big salad eater, I learned that ultrafresh greens have an earthy quality that provides an outstanding backdrop for just-off-the-vine tomatoes. An enormous world of deliciousness was opened up to me, and I could not get enough of it.

Of course, not being that far removed from my post-college poverty and needing to pay for graduate school curbed some of my enthusiasm for discovering all of these magical places. They were still mostly out of my budget for everyday eating, but they did inspire me to find a way to work more local ingredients into my diet. I explored the intimate confines of the All-Local Farmers' Market at 701 Whaley, carefully rationed my City Roots greens to make them last as long as possible, drooled over Facebook photographs of dinner specials at Cellar on Greene and Solstice, and wondered feverishly how soon I would be able to work a community supported agriculture (CSA) share from Pinckney's Produce into my budget (actually, I am still wondering about that).

Thus began my polyamorous love affair with Columbia's local vendors, local ingredients and locally owned restaurants.

Chapter 1

LONG AGO, BUT NOT SO VERY FAR AWAY

In understanding the way food shapes Columbia, it is important to first understand why Columbia exists, how it emerged and how it evolved into the city it is today. Way back in the day, while Charleston was establishing itself as a bustling port town, Columbia was made up of a handful of farms and plantations set up along the banks of two main water supplies, the Congaree and the Wateree Rivers. There was not any kind of centralized market or even a town center, really. It was basically an isolated frontier, and it was fully prepared to stay that way. Then the idea of centrality gripped the fledgling United States, and the fortunes of what are now Columbia and Richland County were irrevocably changed.[1]

When the Revolutionary War came to an end, and the United States was trying to find its footing as its own country, the idea that authority would be more valid if it were concentrated in one location took hold. In South Carolina, it seemed to make sense that the state's governing bodies should relocate from Charleston to a place that that was more centrally located. With the help of pre-Columbia settlers, Richland County and what is now Columbia were created. Shortly thereafter, Columbia became the hotbed of political, cultural and economic life in South Carolina. It took less than fifteen years for Columbia to evolve from unchecked forestlands to the state capital.

The ambitions of Columbia's leaders knew no bounds. As young as the town was, it was already planning to further expand its cultural influence by building what is now the University of South Carolina in an attempt to bring together the otherwise segregated people and cultures of the Upstate, Lowcountry and the Midlands. Created from virtually nothing, Columbia became a vibrant and unique community. As a city, it was "the first in American history to take on the difficult task of transferring the machinery of a functioning bureaucracy to a wilderness setting and making that experiment work."[2]

Part of what made Columbia work, and what still contributes to its successes today, is that city serves as a melting pot of cultures, not just from within South Carolina but from around the country as well. Thanks to the expansion of the university, and the rise of Fort Jackson as the country's largest army training center, Columbia has benefited from the diversity of its population. In 1973, William Harrelson, the South Carolina commissioner of agriculture, described Columbia as "a college town, industrial complex, military base, historical haven, retailing center, and popular 'stopover' for thousands of visitors each year."[3] It took Columbia a long time to achieve the successes it touted in 1973, but in the forty years since then, Columbia has changed some on the surface but not too much at its heart. The city's roots are deep, even if its beginnings were a bit on the humble side.

Like most of America, Columbia's original inhabitants were Native Americans, namely the Cherokees, Congarees, Waterees, Catawbas and Saludas. These tribes not only lent their names to various sites around Richland County but also became the first to engage in trade with the Europeans. The story is the same as it is in the historical annals of colonial America—a handful of men who had settled in a bigger town set off into the wilderness to see if the natives were interested in trading their furs and skins for European novelties.[4]

While few of the early settlers left records of their encounters, John Lawson chronicled his trip from Charleston to the eastern part of North Carolina. Along the way, Lawson stopped in the area that evolved into Columbia, where he was well received by members of the Congaree tribe. He described the group as having a handful of plantations along the riverbank that provided great fodder for their animals. Eventually, the Congarees were pushed a little bit farther west, into what is now Lexington County, as the Wateree tribe took up residence in Richland County.

Neither tribe had been in their new space for long when many of the local tribes decided to make a last-ditch effort to oust the European settlers in the form of the Yamasee War.[5] During the war, most of the local tribes banded together, but the Cherokees kept their own counsel and eventually chose to fight on the side of the Europeans. What resulted were mass casualties, a down-and-out group of colonialists, the migration of a few Native American tribes out of the area and the establishment of a trading factory, run by the government, in an effort to regulate trade between the settlers and the Native Americans. The big winners in all of this turmoil were the Cherokees, whose efforts on the European side of battle did not go unnoticed. As a result, their standing rose in the eyes of the settlers, and this led to increased trade and commerce in what is now Richland County.[6]

As part of their legacy, the Indian tribes who fled during or after the Yamasee War left behind pockets of culture that continue to affect Columbia today. Crops, such as corn and squash, and the techniques used for farming and harvesting, though forgotten for many years, have begun to reappear as current local farmers look for a return to their roots and more sustainable methods of farming. If nothing else, the Native American tribes had a deep reverence for the earth and used methods that were as gentle on the soil and land as possible.

By 1740, Columbia's first permanent nonnative residents had begun to settle in. These first residents were not as wealthy as their Charleston-area counterparts and came mainly from Scotch-Irish, German and English farming backgrounds.[7] Some of the early settlers headed for Columbia, intent on re-creating the rice planting techniques already firmly embedded in the culture in the Charleston area. These individuals established themselves near the banks of the Congaree River in what is now the Lower Richland area of town so that their rice crops could benefit from the area's swamplike ground.[8] Trade routes ran along the banks of both the Congaree and Wateree Rivers, and the land mass in the middle was often bypassed, though this area is what ultimately became the county seat. Gradually, as a result of wars, skirmishes and general conflict on both sides of the Atlantic, more and more people began to flock to South Carolina to establish a new way of life. This resettling brought with it influence from the German and Swiss cultures, British culture and even bits of the more northern realms of the country, as groups from Virginia, Pennsylvania and North Carolina elected to move south instead of into the open frontier of the West.[9]

The start of the Revolutionary War brought with it many changes to South Carolina. An absurd number of battles were fought within the state, and the British retained control of Charleston through 1782.[10] In 1786, as a result of grumblings from the settlers in the Upstate and from the general belief that government should legislate from a central location, Columbia was named as the new state capital. Plans for the city were commissioned and executed, and by 1791, South Carolina's General Assembly was firmly ensconced in the Midlands.[11]

Looking down Main Street to the capitol building, circa 1905. *Library of Congress.*

Columbia slowly began to grow and soon became a hub of culture and industry in South Carolina. First came the opening of South Carolina College (now the University of South Carolina) in 1801, creating an educational center for the state's elite. The invention of the cotton gin made South Carolina the largest manufacturer of cotton in the country. As a result, textile mills, tanneries and iron foundries began to spring up all over the city. The arrival of the railroad and steamboats traveling downriver to Charleston made exporting the city's goods to the rest of the country much easier than it had been.[12]

In the sixty years before tensions came to a head and South Carolina once again found itself at the center of a war, Columbia possessed a certain chameleon-like quality, showcasing different attributes depending on who was looking:

> *Large, prosperous plantations with hundreds of slaves certainly did exist, and its growing urban center possessed much of what any American city of that era could offer—canal travel, railroad tracks, fire companies, telegraph wires, traveling stage shows, circuses, and numerous cultural and educational facilities, as well as an assortment of shops, dry goods emporiums, and saloons. Nearly every visitor was impressed by Columbia's fine homes and broad streets, its intellectual atmosphere, and the pervasive presence of black, some well dressed and well cared for, some not.*[13]

Once again, Columbia had shown itself as a hub of burgeoning prosperity, especially as compared to some of the other states in the country. Unfortunately, these halcyon days were not to last. While the invention of the cotton gin radically increased productivity among South Carolina's cotton farmers, it did the same for the rest of the United States, leading to an oversaturation of the market. South Carolina's economy took a dip, and that, paired with rising dissatisfaction with the federal government, set the stage for the coming war.

Beginning in 1860, when it led the way by seceding from the Union, and not ending until about thirty years later, South Carolina was in turmoil. Columbia, in particular, was hit hard by the devastation of the Civil War and the tumult of the Reconstruction era. John Hammond Moore, when describing the feeling of that war and postwar years in his book, made note of four characteristics that shaped the period: "Exultation, despair, dreams, and hope—all were part of the local scene during these tumultuous years."[14] Columbia was virtually destroyed by the war. On his infamous march to Atlanta, General Sherman set fire to everything in his path, leaving only the University of South Carolina standing, and that was only because it was being used as an army hospital to treat his own soldiers. By the time the war ended, Columbia was in shambles, and it did not seem likely that the city would be able to recover in a short period of time.

Reconstruction did not treat Columbia much better than the Civil War had. The economy was at rock bottom, the male population had been decimated and the city was in ruins. Previously observed social mores and customs were pushed aside as the former policy makers found themselves without money or their previous political clout, and emancipated slaves fought to gain some of that wealth and position. As a largely agrarian community, residents of Columbia

Political unrest in Columbia just after the end of reconstruction, circa 1877. *Library of Congress.*

and Richland County tried to rebuild their lives according to the only methods they had ever known. Cotton farming could not save them, though, as prices were too low to allow for much profit. Farmers and plantation owners were also faced with the reality that much of their workforce was now a community of freedmen. The plantation style of life was dead, and Columbians had a steep learning curve ahead of them.

More than a decade after the end of the Civil War, Columbia finally started to regain its footing. The 1880s ushered in a period of standardization, modernization and technological advancements. Telephones, electricity, mandated public education and the building of a canal along the Congaree River dominated the decade. The canal, in particular, was a source of much rejoicing among Columbia's still struggling residents, who believed that once the canal was in place, hundreds if not thousands of jobs would be created when factories were built on its shores. The canal pipe dream did not pan out as quickly as imagined, but by the turn of the century, the resurgence of cotton mills and factories was in full swing.[15]

The 1890s ushered in a period of industrialization and long-overdue governmental reform. Under Governor Benjamin Tillman, white farmers regained some of the prosperity that had eluded them during the previous two decades. This group regained most of its political clout, while on the other side of the spectrum, blacks became more and more segregated

18

from their white counterparts.[16] Despite Tillman's successful efforts to fully implement Jim Crow laws, his political reforms laid the groundwork for the prosperity that followed as South Carolina ushered in the twentieth century.

The early 1900s were a time of great expansion for Columbia and its surrounding communities. Towns on both sides of the river gained their incorporation, and residents could now claim that they lived in Lexington, Shandon or Elmwood Park. The dreams of the canal bringing more factories and mills to the town were realized as industry in Columbia once again boomed. Banks and more white collar–oriented businesses sprang up along Columbia's main street, and the city's residents swelled with pride at their towering new skyscrapers. By March 1902, Columbia could best be described as a city under construction as anyone with even a remote interest in industry broke ground on a building for his pet project.[17]

Just outside the city limits lay the majority of Richland County, which was still a largely agricultural area. As America kept an eye on the news from Europe and debated whether or not to embroil itself in the First World War, Columbia's civic leaders turned their eyes toward bolstering the agricultural trade of their county. This shift from big business back to farming was due, in part, to fears of a decimated cotton crop as boll weevils moved north from Charleston and fruitless efforts to turn Columbia into a major inland port. County leaders realized that it would be a mistake to ignore a huge revenue source and so set a course to bring agrarianism back into the limelight. South Carolina held the statewide Conference on the Common Good in 1913 to discuss the needs of farmers, their families and their communities. Speakers discussed new types of crops and farming methods. As a result, agriculture was once again at the forefront of the Columbia area's economy. Unfortunately, these monumental inroads were not to last.[18]

During the war, the United States government established Camp Jackson, which later became Fort Jackson, on the outskirts of Columbia. The large influx of new residents also gave a new look to the city. The creation of the army camp brought with it creation of thousands of jobs, further contributing to the economic prosperity of the city. Indeed, the *State* newspaper raved about the benefits of Camp Jackson and the ways in which it was shaping the city:

> *The unvarnished truth is that Columbia, compared with most other Southern towns, was, until lately, poor in money. It had centrality of location, water power, railroad facilities, colleges and schools, and many other advantages, but it is not to be denied that the number of well-to-*

Granby Cotton Mill, circa 1909. *Library of Congress.*

do businessmen was relatively small; hence the raising of a fund for any public enterprise could only be accomplished by severe straining. All that Camp Jackson has changed; scores of merchants and other businessmen now are on solid ground, whereas a little while ago those not cramped in their operations by lack of capital were few and far between.[19]

Columbia was in the midst of an economic boom that lasted well into the 1920s. The city's population was growing steadily, mills and big business were running smoothly and even farmers were reaping some of the benefits. The city continued to diversify itself as soldiers came in to train at Camp Jackson and more students flocked to one of Columbia's six universities. Though the cotton crop suffered, farmers put an emphasis on cultivating new crops to further benefit the city's economy.

As in all other parts of the United States, Columbia ran into tough times as the Great Depression hit. While Columbia was certainly hit hard, it was more fortunate than many other cities because its industrial structure was so diverse.[20] That is not to say that Columbia made it through unscathed. As jobs were lost and wages decreased, Columbia's citizens were not in a good way. But an overwhelming sense of community, still prevalent today, was shown as groups such as the American Legions, Kiwanis and Rotarians coordinated a Christmas campaign in 1930 to collect food, clothing and toys for those who were suffering the most.[21] By the end of 1932, officials estimated that about twelve thousand of Columbia's residents were looking for work. The problem was exacerbated by the sheer fact of Columbia's location on two major railroad lines. Drifters looking for work often ended up in Columbia, draining the city's meager resources as they were housed in the county jail, at Camp Jackson or at the local Young Men's Christian Association (YMCA). Finally, in the middle of 1932, the federal government came to the rescue, as President Hoover initiated small programs to help alleviate some of the burden on local governments. It was not until President Roosevelt launched his New Deal policies that things really began to turn around.[22]

Just as Columbia was pulling itself out of yet another tumultuous period in its history, things took a turn for the better, as Camp Jackson was renamed Fort Jackson and took on the role of training army infantry recruits for World War II. The influx of soldiers and their families raised Columbia's population by nearly one-third and greatly contributed to the further diversification of the city.[23]

The end of World War II ushered in a period of nationwide uniformity, with advances in technology, farming practices and all-around prosperity—all with civil rights issues simmering just under the surface. Columbia was not immune to these changes, but its residents still managed to maintain ties to their roots, their pasts and their traditions. Wrote John Moore in 1993:

> *Unlike most Americans, Midlands residents can drink countless glasses of iced tea during any season of the year, eat grits for breakfast day in and day out, and feast on barbecue and fried chicken several times a week without complaining. Whether white or black, they probably will have some familiarity with their forebears and brag of ties to a small-town, rural way of life a few generations back. And they will be conscious of their Midlands heritage, caught as they are between upcountry and low.*[24]

Even now, twenty years later, his statement still holds true. Columbia residents, whose families have been residents for generations, are extraordinarily proud of their heritage. It could come from a sense of pride in being a unique community within the state, since Columbia was settled by people who worked hard for what they earned. Or it could be that they are proud of the fact that in times of crisis, Columbians have always banded together to support one another. It may even be due to the fact that Columbia's residents have great pride in their city, knowing it to be so deeply entrenched in the cultural, political, educational and economic happenings within the state.

Today, Columbia is well on its way to recovering from the latest crisis it has faced, with the economic recession beginning in 2008. Mayor Steve Benjamin is on a mission to revitalize the city's Main Street area, which was a hub of activity and growth in the 1970s and 1980s. So far, his plans appear to be working with the opening of Mast General Store,

Main Street in Columbia, circa 1901. *Library of Congress.*

Paradise Ice, the Oak Table and an expansion of popular Five Points coffee shop Drip. The 2012 move and subsequent transformation of the All-Local Farmers' Market into Soda City has made Saturday mornings on Main Street a premier event. Columbia's unemployment rate was at 7.6 percent in December 2012,[25] showing a decrease from 8.9 percent in November 2012.[26]

With a new spin on centuries-old techniques, City Roots, Anson Mills and Caw Caw Creek are bringing farming and animal husbandry back into style with a vengeance. The chefs at restaurants like Terra, Rosso, Motor Supply Company and Solstice are using their food to remind Columbia residents just how rich their local food systems are. Small neighborhood markets like Rosewood Market and even corporate giant Whole Foods are providing stages for locally grown, raised or manufactured products to shine. Small family farms are bringing the world of sustainable eating to Columbia residents' fingertips through their offered CSA shares and their presence at Soda City and the Vista Marketplace. Columbia's series of successful food and culture festivals each year showcases the wealth of options that the city and its surrounding regions have to offer. Each of these outlets creates the community spirit for which Columbia is so well known.

From its very beginnings, Columbia has been something special. It was "born of compromise and has spent much of its life balancing views of contending forces—rural and urban, farm and factory, Piedmont and coastal."[27] This enduring pragmatism and balance is what led Columbia throughout all of the turbulent times in its storied history. From the Revolution to the Civil War, its devastating aftermath and the modern era, Columbia has persevered, thanks in large part to the dedication of its citizens and the tight-knit community that it has created.

Chapter 2

GOING TO CAROLINA, WON'T BE LONG 'TILL I'LL BE THERE

A s the capital city of a state with a strong agrarian background, it should not come as a surprise that Columbia has fully embraced the local and sustainable food movement currently sweeping the nation. It could be argued that South Carolina has always embraced the tenets of both the local food and sustainability movements. Since the establishment of South Carolina in 1670, agriculture has been the driving force of the economy in the state, with an annual impact of about $34 billion.[1]

In its earliest days, South Carolina prospered thanks in large part to the fertility of the land in the Lowcountry and the harbor in what is now Charleston. Trade and industry spread as the number of settlers increased and started to migrate outward from Charleston. Rice became a huge staple crop as people settled along the swampy marshlands, and slaves from the rice growing regions of west Africa brought their farming skills and techniques to the area. As a result, in the 1700s, most of South Carolina's economy revolved around rice cultivation, due in large part to the high prices the crop garnered when exported to England.[2] Thanks to the slaves' knowledge of rice cultivation, harvesting methods remained similar to those they had used in Africa. When it came time to harvest, "women processed the rice by pounding it in large wooden mortars and pestles, virtually identical to those used in West Africa, and then 'fanning' the rice in large round winnowing baskets to separate the grain and chaff."[3]

As the eighteenth century wore on, several varieties of rice began to develop, the most prominent of which was the variety known as Carolina

Gold. The production and exportation of Carolina Gold was so successful that in Europe and other parts of the United States, it was referred to as the best rice in the world. But the prosperity was not to last. Like most of the Southern states, South Carolina's world was turned upside down with the coming of the Civil War. The devastation resulting from four years of war took its toll, and South Carolina was forced to start from scratch. For a time, the rice crop continued to thrive, but production eventually faltered, and by the time the Great Depression rolled around, Carolina Gold Rice had ceased to be a major economic contributor as the state embraced more efficient and higher-yielding rice varieties.[4]

In recent years, however, a handful of South Carolinians have dedicated themselves to bringing Carolina Gold back into the spotlight. There is even a Carolina Gold Rice Foundation, whose mission is to "advance the sustainable restoration and preservation of Carolina Gold Rice and other heirloom grains, raise public awareness of the importance of historic ricelands and heirloom agriculture, encourage, support, and promote educational and research activities focused on heirloom grains, and serve as an information resource center to provide authentic documentation on heirloom grain culture and heritage."[5] Campbell Coxe and his plantation in Darlington County can take credit for reawakening interest in one of South Carolina's oldest and most historic crops. Using the Della variety of rice seed, Coxe's crops have found success in the Charleston food market. Donald Barickman, one of the premier chefs in the Lowcountry, is one of Coxe's biggest supporters. In an interview with *Discover Charleston*, Barickman was quoted as saying, "When I met the Coxes at a wedding in Knoxville, they told me how they were bringing this old heirloom rice variety back to life on their plantation on the Great Pee Dee River. They offered me some of their rice to play around with. Ever since then, there's really been no other rice for me."[6]

In Columbia, located in a part of the state never really known for its rice growing, Carolina Gold is making a comeback, thanks to the folks at Anson Mills. Using organic and sustainable growing practices and a variety of heirloom seeds, Glenn Roberts and the rice, grits and grains he produces at Anson Mills have taken America's culinary world by storm.[7]

Head judge on *Top Chef* and renowned New York City chef and restaurateur Tom Colicchio is a big fan of Anson Mills. In 2009, Colicchio spent a day with Roberts and a few of his colleagues to learn all about what Anson Mills had to offer. Colicchio was especially impressed by the process, the variety of products, the story behind it all and, in particular, the Perla Bianca corn kernels that Roberts had just obtained. During his visit, Roberts milled some of the just received kernels into polenta, and Colicchio was bowled over. "One whiff of the freshly milled corn and it's obvious why Glenn's fighting for it. Its floral, milky scent and sweet taste are unlike anything I've known," he wrote in a blog entry about his visit.[8] Perla Bianca is not the only distinct crop that Roberts cultivates at Anson Mills. Specializing in heirloom varieties of grains, rice and corn that are parts of a long South Carolina tradition, the idea for Anson Mills was inspired by the idea of the "Carolina Rice Kitchen," a phrase used to describe the nineteenth-century South Carolina food culture that used rice instead of other grains as its primary ingredient.[9]

Traditionally, a rice kitchen has been linked to an almost deity-like worship of rice. In cultures where rice was the only staple, it was highly revered and often looked on as a giver of life. A more modern view, and indeed the one that seems most compatible with the idea of the Carolina Rice Kitchen, is that rice is served with every meal and treated with respect. Throughout most of the peak rice producing period in the Lowcountry, rice was a part of every meal, despite the availability of other substitute grains.[10] In South Carolina, the traditional rice kitchen brought together flavors and influences from three primary cultures—the African slaves, the Native Americans and the Europeans who orchestrated the necessary irrigation methods—to create a cuisine that was unique to the region.[11]

Because of the times, the products made possible through the Carolina Rice Kitchen were ones that would make modern chefs sing. Using sustainable methods of farming, put into practice based on the rice farming knowledge of the slaves, Carolina Rice Kitchen cuisine is the poster child for local and sustainable eating. At Anson Mills, those techniques and traditions are a large part of why it operates the way it does. Using the types of heirloom grains common in antebellum households and the agricultural techniques specific to the time, Anson Mills hopes to be able to contribute to the resurgence of the Carolina Rice Kitchen. After all, the cuisine from the rice kitchen's heyday was dependent on "a complex agricultural system suited to local conditions and cultural needs."[12] This kind of thinking is in line with what many in Columbia hope to achieve in the local food movement as well. As a result, grains and rice from Anson Mills have found their way into dishes at prominent local restaurants like Rosso. Anson Mills also sells its products at Soda City for everyday use. Anson Mills' commitment to traditional farming methods, sustainable techniques and high flavor is what has made its foray into the rice world so successful.

Chapter 3

WELCOME TO FAMOUSLY HOT COLUMBIA, SOUTH CAROLINA

Columbia, South Carolina—the home of the University of South Carolina, the state government and just over 130,000 people (according to the July 2011 census)—is currently undergoing an existential crisis. As a city in proximity to more well-known locations like Atlanta, Charlotte, Charleston and Savannah, Columbia's many charms are often lost on nonresidents. In 2008, the Columbia Visitors' Bureau set out on a mission to change that—first by changing the city's slogan from "A Capital Place to Be" to "Famously Hot." Anyone who has spent any time in Columbia knows that calling the city "famously hot" is well deserved. With average temperatures in the mid- to upper nineties and an average humidity of 973 percent (not really, obviously, but the humidity in Columbia in the summer is brutal), the city certainly matches the definition of hot. But the temperature is not the only thing making Columbia sizzle. There are some extraordinary things going on in this town that deserve recognition. From the cultural festivals that commandeer the streets in the Vista on most weekends in the fall to unprecedented success from the university's football team to the giant strides being made in local food, Columbia's residents finally have something to brag about.

Food-wise, Columbia is called home by several renowned chefs who are putting their mark on the food industry by showcasing local ingredients in their restaurants. The city has seen a resurgence in farming, with the opening of City Roots and by others on the outskirts of town. The demand for local food is skyrocketing, as evidenced by the increase in farmers' markets and

CSA shares. The slow food movement is taking the city by storm, with Slow Food Columbia continuing to grow. Partnerships between local chefs and local farmers have resulted in an increased access to and interest in farm-to-table eating, clean eating and locally grown products. Columbia is starting to make a name for itself on the national scene, and it is positioning itself to compete with destinations like Charleston and Atlanta.

It has, however, been a long journey to this point. Columbia got its start, purely by chance, when state legislators wanted to move the state capital from Charleston. Plans for the city were commissioned, making Columbia the second planned city in the United States, and the population rapidly increased, boasting about one thousand residents by the start of the nineteenth century.[1] Because of its centralized location, Columbia was an excellent choice for the state capital. Thanks to the doings of the legislature and the opening of South Carolina College, which later became the University of South Carolina, in 1801, Columbia became the hub of political, educational and commercial activity early in its history.[2]

Prosperity continued as the cotton trade grew and technologies advanced. By the start of the Civil War, Columbia was a fully modernized city, with more than eight thousand residents. Unfortunately, the city's prosperity was not to last. The Civil War decimated the city, as General Sherman's troops marched through, burning everything in their path. It took some time, but eventually Columbia was on the verge of recovery. It is odd to think that Columbia owes both its near downfall and its remarkable turnaround to war, but that is how it happened. As the United States found itself embroiled in World War I, the government set up a training camp on the site of what is now Fort Jackson. Thanks to a new sense of national importance that only grew as improvements were made to the army base in response to World War II, Columbia was able to pick itself back up, dust off the ashes and emerge as a diverse southern city.[3]

That type of diversity is evident when reading a list of Columbia's signature events on the Columbia Visitors' Bureau website. Events such as Tartan Day South, Main Street Latin Festival, Columbia's Greek Festival, Columbia Italian Festival and SC Pride Parade and Festival showcase the diversity in culture that makes Columbia so unique.[4] Cultural diversity is not the only thing that makes the capital city so special. The list of events revolving around food is even more extensive than the list of events celebrating cultural heritage. All year, Columbia showcases its burgeoning food scene, with participation in the ever popular Restaurant Week, the Five Points Chili Cook Off, the Rosewood Crawfish Festival, the Irmo

Okra Strut, the Lexington County Peach Festival, the Palmetto Tasty Tomato Festival, Viva La Vista and the South Carolina Oyster Festival.[5]

Restaurant Week offers diners a chance to check out some of Columbia's best restaurants for a fraction of the cost. Most participants offer a full three-course meal, and some even offer a complimentary glass of wine. Establishments specializing in the use of local and sustainable ingredients are some of the most common participants, providing patrons the chance to not only try a new (to them) restaurant but also get a feel for some of the wonderful local vendors Columbia has to offer.

The Five Points Association hosts a chili cook-off each fall, primarily as a fundraiser but also to see which local residents can make the best chili. Some participants use tried-and-true family recipes, and some try something brand new. During the cook-off, the Five Points Association encourages other Five Points merchants to showcase their own products.[6] Five Points has always been a slightly quirkier section of Columbia where local entrepreneurship has been able to thrive. Supporting these merchants helps create that sense of community of which locavorism so proudly boasts.

The Rosewood Crawfish Festival is one of the premier events in Columbia. Held annually in the Rosewood section of the city, the festival imports thousands of pounds of fresh Louisiana crawfish and then celebrates their deliciousness. The festival also brings in local vendors to supplement the crawfish and to cater to those with a taste for something a little bit different. Since a large section of Rosewood Drive is closed off to accommodate the festival, organizers found a way to take up more of the space. They invited local artisans and craftsmen to showcase their products.[7] In a way, the crawfish festival can be thought of as a miniature farmers' market—between the food selections, the local goods and the community spirit surrounding it, the festival captures that same sort of farmers' market vibe.

Nothing screams *local* much more than an entire festival revolving around a local crop. But that is exactly what the Irmo Okra Strut is. Billed as "the nation's original celebration of okra,"[8] the okra strut has been going strong since 1973, when it was implemented as a fundraiser for the Lake Murray–Irmo Woman's Club. Since then, the festival has grown, now running for two days, and features such community-building events as a craft fair, food from local restaurants, an okra eating contest and, of course, live enterntainment.[9]

Leave the Columbia city limits and head across the river, and you will be in Lexington County, one of the Columbia area's more rural locations. Every year, Lexington County hosts a peach festival to showcase one of the state's most popular fruits: the South Carolina peach. Lexington County takes peach appreciation to a whole new level during the festival, as its people incorporate peach pageants, a parade and a recipe contest. Of course, they also include picnic food, complete with a list of peach-based desserts that makes Bubba's litany of shrimp recipes in *Forrest Gump* seem minimal.[10]

The Palmetto Tasty Tomato Festival is a collaborative effort between three Columbia organizations: Slow Food Columbia, Sustainable Midlands and City Roots. Their mission? "To highlight the taste of sustainably grown, local heirloom tomatoes; increase awareness of the biodiversity of foods; celebrate eco-conscious chefs, farmers and backyard gardeners; and assemble our wonderful progressive food community in a fun, family-friendly setting."[11] Really there could be no better place in town to hold an event with such an incredible focus on a local and sustainable product than at City Roots, the city's most visible urban farm.

Viva La Vista is another chance for Columbia to showcase one of its premier neighborhoods. This festival is a daylong affair that provides attendees with the chance to catch some of Columbia's best local bands and explore offerings from local vendors. Most importantly, though, Viva La Vista is a chance to experience an array of different cuisines from some of Columbia's most innovative restaurants. Billed as an "epicurean event,"[12] the 2012 version of Viva La Vista featured food and drink selections from an array of Vista-based establishments including Blue Marlin, Motor Supply Company, Flying Saucer and Cupcake. With so many different options, there was definitely something there to entice any kind of palate.

The South Carolina Oyster Festival celebrates these pearls of the sea by steaming and frying more than ten thousand pounds of fresh oysters. Like many of Columbia's festivals, this one is held outdoors, even though it happens in November, to take advantage of Columbia's relatively mild climate. At this festival, if you are still able to move after stuffing yourself with oysters, you can expect live music, local artisans and craftsmen, as well as an array of food and drink options. To further add to the local feel for the event, it is held at the historic Robert Mills House, right in the heart of downtown Columbia.[13]

Late in 2011, another series of events began that served to supplement Columbia's traditional festivals and put an even bigger spotlight on the importance of bolstering the local community and committing to sustainable eating. The Harvest Dinner Series brought together Ryan Whittaker, executive chef at @116 Espresso and Wine Bar, the staff at City Roots and Vanessa Driscoll, mastermind behind the Southern Greenie (an event planning firm specializing in green events). Each dinner is held on the farm at City Roots. Whittaker plans the event's menu based solely around what produce can be picked that day from City Roots and what proteins are available from other partnering farms and then brings in a guest chef from one of Columbia's star restaurants to make one course. Each dinner starts with a cocktail hour, when guests are invited to tour the farm and ask questions. Cocktails are made with organic spirits from American Harvest and are mixed with juices made from fresh seasonal fruit. When it is time to eat, guests gather at a large communal table to enjoy the meal and have either a sustainable beer or a sustainable glass of wine. Both the chefs and the farm staff join in the meal to answer questions about the farm at large or about specific menu items. Once dinner has concluded, guests have a chance to purchase produce from the farm or to place an order for the beer or wine served with dinner. On evenings when it gets a little too chilly to eat outside, dinner is served in the City Roots greenhouse, creating an even more unique experience.[14]

All of these wonderful events are going on, showcasing the best of what Columbia has to offer, both as a cultural center and as budding food community. But how do you get involved? One way would be to become a part of Slow Food Columbia, the capital city's branch of the national Slow Food USA organization that comprises more than two hundred chapters. Its mission, as stated on its website, is to "support the movement behind GOOD, CLEAN and FAIR foodways in the Midlands and beyond. Our convivium hosts workshops, potlucks and other events throughout the year to celebrate local and seasonal flavors; to showcase the culinary talents of our region's chefs, farmers, and artisan producers; to strengthen connections between members of our local food community; and to educate the public about the importance of knowing where your food comes from."[15] As an extension of the national organization, Slow Food Columbia seeks to promote the link between a sense of community, environmental consciousness and the food that people use to nourish their bodies.

Both Slow Food USA and Slow Food Columbia members strive to implement good, clean and fair principles into their lives. Each of these terms has its own common connotation, but for Slow Food, the definitions are a bit expanded. "Good" refers to the use of healthy animals and fresh plants to create delicious food. Celebrating this goodness also means acknowledging and reveling in cultural, ethnic and regional diversity. In this case, "clean" refers to clean eating, where food comes from places that adhere to sustainable agricultural methods in order to create a positive impact on the environment. "Fair" tackles the idea of food equity. Slow

Food organizations firmly believe that good, clean food should be accessible to everyone, regardless of socioeconomic status. "Fair" also refers to the division of labor and the idea that those who work to produce the food should be adequately compensated and treated fairly.[16]

The list of partners and supporters that Slow Food Columbia features on its website reads like a who's who of the Columbia food scene. Featuring restaurants, farms and farmers, local bloggers, regional magazines and assorted sustainable organizations from the Columbia area, it is pretty clear that Slow Food Columbia has made great strides in uniting the Columbia food community. Slow Food Columbia also has a hand in several food-related events around the Midlands, including the Palmetto Tasty Tomato Festival and Slow Food at Indie Grits, which is a partnership with another Columbia institution, the Nickelodeon. With its focus on local and sustainable eating, environmental activism and involvement in the community, Slow Food Columbia is a prime example of what it means to be part of the local food movement.

Chapter 4

THE ADVENT OF A LIFESTYLE

On a national scale, eating locally is not really a new thing. For years, cities whose names conjure up images of specialty cuisine have been focusing on local ingredients in their dishes. New York, San Francisco, Seattle and even Charleston have become meccas to foodies seeking incredible meals that showcase ingredients local to the area. Environmental concerns, a desire to return to a simpler way of life and skyrocketing interest in healthy eating and nutrition are just a few reasons why people are flocking to establishments intent on providing the best local fare around.

So what exactly constitutes "local"? Really, it is one of those terms that is subject to interpretation, based on your personal feelings. In the United States, about two-thirds of people consider food to be local if it comes from within a one-hundred-mile radius of home, according to a 2008 survey by the Leopold Institute. This definition would certainly resonate with Sage Van Wing, who along with three of her friends coined the term "locavore" when she decided to try local eating after reading Gary Paul Nabhan's account of his own local eating experiment, *Coming Home to Eat*.[1]

Using the one-hundred-mile definition is an easy way to determine whether or not food is local. Limiting eating habits to within your own foodshed is another. A foodshed, like a watershed, shows the flow of food from a specific area. Being familiar with your local foodshed will allow you to know exactly what kind of journey your food took to get to your dinner table.[2]

While distance may be the primary criterion used for determining what constitutes local food, another facet that locavores tend to consider is sustainability—in other words, what impact is your food having on the environment? Another one of those foodie buzzwords is sustainable agriculture, which "encourages using renewable resources to increase farming profitability and improve environmental and socioeconomic health."[3] Because sustainable farms contribute much to the local community, their practices start the beginning of a cycle in which restaurants and foodies alike purchase their products. With a high demand and consistent sources of revenue, farms are able to produce more, which leads to additional sales and additional exposure within the community. In essence, locavorism is a lifestyle, not just limited to food but also consisting of an active interest in the community.

Californians might take credit for kicking off an interest in locavorism, but the lifestyle has quickly spread throughout the country. And before you claim that this is clearly a movement supported by old hippies, vegans, Brooklyn hipsters and the 1 percent crowd who can afford to pay extra for organic food, just know that locavorism is accessible to everyone. All you need to do is to take a trip to a local farmers' market. There you will find locally grown and produced sustainable foods (some even may be certified organic) at affordable prices, or at least at prices comparable to some of the more processed items you can find at the grocery store. In California, part of the state's SNAP (Supplemental Nutrition Assistance Program, or food stamps in the vernacular) benefits include vouchers for local farmers' markets, specifically to provide access to healthier food options to California residents.[4] The same holds true in South Carolina, where SNAP recipients can use their benefits to purchase fresh produce items from authorized vendors.[5]

It is no secret that the United States has an obesity problem. From First Lady Michelle Obama's Let's Move! Campaign to television shows like *The Biggest Loser*, the focus on healthy dietary options is sharpening. Part of that is due to the spread of the locavore movement. In 2007, more than one thousand school districts in thirty-five states across the country started incorporating local foods into their school-provided lunches. Now, the National Farm to School Network has representatives in all fifty states who work directly with schools to connect them with local farms with the intent of "serving healthy meals in school cafeterias, improving student nutrition, providing agriculture, health and nutrition education opportunities, and supporting local and regional farmers."[6]

Founded in 2007, shortly after Californians set the country talking with their focus on local eating, the National Farm to School Network is on a mission to

Strengthen children's and communities' knowledge about, and attitudes toward, agriculture, food, nutrition and the environment; Increase children's

participation in the school meals program and consumption of fruits and vegetables, thereby improving childhood nutrition, reducing hunger, and preventing obesity and obesity-related diseases; Benefit school food budgets, after start-up, if planning and menu choices are made consistent with seasonal availability of fresh and minimally processed whole foods; Support economic development across numerous sectors and promote job creation; Increase market opportunities for farmers, fishers, ranchers, food processors and food manufacturers; Decrease the distance between producers and consumers, thus promoting food security while reducing emissions of greenhouse gases and reliance on oil.[7]

In short, the network is working with schools to instill a love of local and sustainable methods into the minds of the next generation. Relating directly to the mission of the Let's Move! Campaign, the National Farm to School Network was recognized in a White House Task Force report as a strategy for combatting and preventing childhood obesity.

South Carolina is participating in the Farm to School program thanks to a joint effort between the state's Department of Education, Department of Agriculture, Department of Health and Environmental Control and Clemson University's Youth Learning Institute. In 2012, this program was part of a two-year pilot study. However, there are still eight districts and thirty-six schools across the state that are participating in the program.[8]

Thanks to grants from the Center for Disease Control, the nation's children have access to information about quality locally grown food. While that is certainly a wonderful trend and a program that should continue—and will, if the increased participation over the last five years is any indication—the country may find itself with better-educated children than parents, at least when it comes to eating locally. So how can the parents keep up? By doing their own research into locavorism and finding out what style works for them and their families.

Much like vegetarianism, there are a few different styles of locavorism: ultrastrict, Marco Polo and wildcard. Ultrastrict locavores determine what they consider to be their local radius and, as their name implies, stick strictly to products produced and grown within that area. That means they often forego things that most of us find we cannot live without, like coffee, olive oil, chocolate, beer and even many spices. In vegetarian terms, ultrastrict locavores are the equivalent of ardent vegans. Marco Polo locavores are a bit less strict, but not by much. Essentially, they take the ultrastrict rules and bend them to include dried spices because these were items that sailors could carry with them at sea. Finally, there are the wildcard locavores. These are the equivalent of those folks who refer to themselves as vegetarians but still eat seafood. For the wildcards, the focus remains on local products, but they are not as militant in their devotion. Coffee, bananas and olive oil are fair game, though the wildcards may try to justify their consumption of such items by only buying organic or, in the case of coffee, beans marked as fair trade. Wildcard locavorism is the most accessible and, as such, the most common in the local food movement.[9]

So why in the world would you even entertain the idea of becoming a locavore? To start, as you can see from the instance of the wildcard locavores, locavorism is what you make it. It is a lifestyle that allows you to pick and choose which parts you will embrace. You have the option to be as strict or as flexible as you wish. You can even go the extremist route and become a yokelvore, where you eat as locally as possible by only eating food from your own backyard. Alternatively, you can make an effort to consume as many local products as possible but also acknowledge that you may make exceptions for items not native to your area. Regardless of which type of locavorism you choose to follow, there are several additional benefits. One of the primary reasons people seek out a locavore lifestyle is to better their health. Clean eating makes for a healthier diet, which in turn leads to fewer health problems. Some people also do it for environmental reasons. Local and sustainable farming methods help decrease greenhouses gases and allow individuals to lessen their own carbon footprint. Finally, turning to a locavore lifestyle supports the local economy. In addition, this type of lifestyle allows locavores to develop a more personal relationship with their communities and to develop a deeper appreciation of the food they consume.[10]

Still not sold? Jennifer Maiser, in a column originally published on her blog "Life Begins at 30" in 2005, outlined ten reasons for eating locally: "Eating local means more for the local economy…Locally grown produce is fresher…Local food just plain tastes better…Locally grown fruits and vegetables have longer to ripen…Eating local is better for air quality and pollution than eating organic…Buying local food keeps us in touch with the seasons…Buying locally grown food is fodder for a wonderful story…Eating local protects us from bio-terrorism…Local food translates to more variety…Supporting local providers supports responsible land development."[11] In her post, Maiser elaborated on each of these points, providing detailed explanations to support her opinion. Some of it cannot be disputed—local food tasting better and being fresher? Most definitely. Tomatoes, cucumbers and peppers straight from the garden will beat out grocery store produce any day of the week.

Chapter 5

REDUCE, REUSE, RECYCLE

Another one of those buzzwords that you hear, often mentioned in the same breath as "local food" or "locavore," is "sustainable" or some derivation thereof. Sustainable can be defined as being "of, relating to, or being a method of harvesting or using a resource so that the resource is not depleted or permanently damaged."[1] When applying that definition to farming and other agricultural pursuits, it is not too difficult to see why sustainability has become such a huge issue in the foodie world.

But why is it important? Aside from the fact that using sustainable harvesting methods helps do things like keep animal populations from being depleted (ever notice that wild salmon only comes from the West Coast? That is because salmon from the East Coast have rapidly diminishing populations), there are all kinds of other ecological and economic factors to be considered.

The technological advances in farming techniques since the end of World War II have fueled the mass production, reduced labor and the associated costs and have generally made food more accessible. Unfortunately, at the same time, these advances have wreaked havoc on both the land and the economic prosperity of small farms and rural communities. It has really only been within the last twenty years that research into the best practices in agriculture has changed direction to focus more on how to if not replenish the land, at least change techniques for the future.

Researchers at the University of California–Davis have been spearheading these initiatives through their work at UC-Davis's Agricultural Sustainability

Institute. In the course of their research, they have identified three primary goals of sustainable agriculture: environmental health, economic profitability and social and economic equity. As a consumer, and ideally one interested in local food, these goals should be easy to relate to the locavore mentality, particularly those concerning profitability and economic equity. The environmental impacts are also important to those desiring a locavore lifestyle, as sustainable farming methods directly affect the local community.

As the UC-Davis team has noted, "A *systems perspective* is essential to understanding sustainability. The system is envisioned in its broadest sense, from the individual farm, to the local ecosystem, *and* to communities affected by this farming system both locally and globally. An emphasis on the system allows a larger and more thorough view of the consequences of farming practices on both human communities and the environment. A systems approach gives us the tools to explore the interconnections between farming and other aspects of our environment."[2] The integration of environmental issues, economic issues and community issues is the key to a successful sustainability strategy. In a nutshell, every member of the community has a part to play in sustainable agriculture. Not just the farmers, but the policy makers in the government, the consumers and the retailers as well. To successfully support a sustainable culture, each group needs to support the others.

In South Carolina, and Columbia in particular, the organization Sustainable Midlands exists to make sure that this happens. Its mission is to "advocate, educate, and celebrate solutions that balance the needs of the community, the environment, and the economy."[3] It does this by providing research and sponsoring events to raise awareness and interest in sustainability issues in the Midlands. It partners with a local school district to help teach local students about the importance of environmental health and some of the environmental issues they will face when they grow up. Initiatives relating to clean water and sustainable communities, as well as providing guidance and tips to local businesses who wish to become more conscious of their impact, have kept this organization running and thriving, even within the urban confines of Columbia.

One of the largest population segments in Columbia is the student body at the University of South Carolina. With a total population surpassing thirty thousand students, USC has to exert an enormous effort to minimize its carbon footprint. Luckily, students, faculty and staff can turn to Sustainable Carolina for inspiration. Much like Sustainable Midlands, Sustainable Carolina has a mission to "educate and transform the campus and community by promoting collaborative relationships among students, faculty, staff, and community members for exploring and implementing the changes required to create a sustainable campus and society."[4] By partnering with several offices and organizations across campus, Sustainable Carolina is already leaving its (very faint and not at all environmentally unsound) footprint on the university.

In the group's first annual report from the 2011–12 academic year, university president Harris Pastides commended Sustainable Carolina for taking on a leadership role in the campus community to bring sustainability issues to the forefront of the university's agenda. Dr. Pastides continued by outlining strategies that USC would implement in the future to progress toward the ultimate goal of carbon neutrality. As Dr. Pastides noted, "If we do nothing, the world we leave to the generations that follow us will be mired in conflict over resource allocation and will be depleted of its natural beauty."[5] Though Dr. Pastides's remarks were in reference to the construction of energy-efficient buildings across campus, they can easily be molded to fit the food landscape as well. In fact, one of Sustainable Carolina's biggest accomplishments was the receipt of a grant from the Environmental Protection Agency (EPA) to start a pre-consumer compost program. As the project suggests, the students involved take food waste and combine it with other campus yard waste. The resulting compost is used in landscaping efforts across campus. Though the composting program only began during the 2011–12 academic year, there are big plans for expansion and continuation in the coming years.

So you see, sustainability is not just about agricultural practices, though the agricultural aspect is quite a large one. Sustainability is also about environmental impact, whether through growing techniques, harvesting methods, building methods or transportation methods. With all of these different facets to sustainability, it should come as no surprise that the term is often used right alongside "local" in the food world. For those locavores subscribing to the one-hundred-mile radius philosophy, "food miles," or the distance food has to travel to get to your table, have a huge impact on sustainability. Food grown within that one-hundred-mile limit does not need to travel as far, which implies that there are fewer carbon emissions and, in turn, less of a negative impact on the environment.

Chapter 6

HOW DID WE GET HERE?

W hy the resurgence of local food? Has this been a long time coming? Is it, like its much-touted practices, sustainable in the long term? Is it a national movement or just concentrated in more health-conscious geographic regions? The answer to the middle two questions is a resounding yes.

An emphasis on local food has been growing across the United States for many years. As environmental issues have begun to take center stage in the political arena, farming practices have come under intense scrutiny. Large-scale advances in farming technology have taken their toll on the land, and there is some concern over how the soil will react. Changing methods to incorporate accepted sustainability practices have taken the country by storm. It is a movement that has been sweeping the nation since the early part of the twenty-first century, and it is not just concentrated in certain regions.

As food prices have gone up and the economy has declined, people have started to look to alternative methods of food production—some have started their own gardens, and some have made a commitment to do their part to support the local economy by purchasing as many products and food items from small vendors, who are usually locally owned and locally operated. Prepackaged foods with far more ingredients than necessary have become the norm in the country's grocery store chains. As more and more reports of health and wellness issues trickle through the news, the country's citizens have pushed back by demanding and exploring new ways to obtain fresher and more natural food. All of these factors have led to an increased emphasis on the importance of local food and sustainable farming practices.

As a direct result of large-scale farming practices and mass production of food, Americans have lost the connection to food sources, often choosing convenience over quality. The heart of the local food movement is an interest in protecting the environment and supporting small, local famers whose profits then feed back into the local economy, strengthening society as a whole. Residual benefits often result from supporting local food, producing a deeper connection to the earth, increasing food accessibility and freshness, enjoying better overall health and creating a more sustainable way of living.[1]

In the discussion of local food, there are always a few terms and common questions that arise. You might hear phrases like food system, local food and sustainable practices. Each has its own importance, and each informs the understanding of the other. Food systems are where it all starts. Everything else comes out of a successful food system. There are two types of food systems: global industrial and sustainable local. The first has only one type, with a huge geographic reach. The second has many varieties, as it is concerned with practices and food in multiple regions or locales. The two main components that then break down into further categories are food production and food distribution. Food production refers to the way in which crops or animals are raised, harvested and slaughtered and the way the end result is prepared for purchase, while food distribution is concerned with how food is transported and where and how it is sold.[2]

Food production is the primary concern associated with sustainability. Sustainability issues are often highly politicized because many of the practices that have brought the concept of sustainable farming into the limelight are direct results of mandates and policies handed down from the United States Department of Agriculture and assorted presidents. The same federal policies prevalent in agriculture today first emerged in the 1930s. Even at that time, there were disputes about how farming should progress. In one corner was the Muhammad Ali of this fight: those supporting the basic ideology surrounding agrarian practices that supported and protected small family farms and kept the farmers tied to the land, no matter what the cost. In the opposite corner was the metaphorical Joe Frazier: the notion of progress, innovation and modernization.[3]

Despite the tensions between the two conflicting opinions, the United States was smack in the middle of the Great Depression, and President Franklin Roosevelt felt that something needed to be done to bolster the nation's economy. President Roosevelt was a staunch believer that the economic well-being of rural farm communities was what really drove the economic well-being of the country as a whole. As a result, the president pushed the

Agricultural Adjustment Act through Congress in 1933, creating what has become the modern farm commodity and price support system. At the time, the goal was to create a safety net for the farming community in case things started to go even further downhill during the Depression. Another part of Roosevelt's famous New Deal plan was put into action by the Farm Security Administration. This program was a bit more radical and sought to support financially struggling farmers by improving conservation methods and creating rural rehabilitation colonies. Unfortunately for most farmers, the two different acts created more of a problem than there was initially. The Agricultural Adjustment Act increased the success and profitability of commercial and big business–type farmers, which served to further impoverish the small family farms that the Farm Security Administration was on a mission to protect.[4]

Just as the country was digging its way out of the economic uncertainty of the 1930s, things spiraled out of control in Europe, and all too soon, the United States found itself in the middle of the Second World War. While on a worldwide scale, the war was a tragic and horrific period, in the United States it actually served as a catalyst for increased productivity in the farming community. Advances in farming techniques, technologies and equipment allowed farmers to produce more, which was a necessity in order to support the war effort. As a result, the price system enacted through the Agricultural Adjustment Act, intended to only be a temporary measure, became a permanent fixture in the world of American farming. World War II served to bring about the end of the New Deal measures, and as a result, many of the smaller agencies that President Roosevelt had established fell by the wayside, including the Farm Security Administration. With no governmental programs to continue to protect their interests, many small farmers were left out in the cold as the large-scale farming operations triumphed.[5]

Though it seemed like progress had won out over tradition, there was still much dissension between the two schools of thought within the government. In particular, the postwar secretary of agriculture, Charles Brannan, was concerned that the trends toward bigger farming operations and the innovations in technology were detrimental to the lifestyles of small farmers. In 1951, Brannan commissioned a review of the Department of Agriculture's policies to see if his suspicions were justified. Brannan was a big believer in the preservation of traditional farming methods because he thought that traditional family farms were the backbone of America. At this point in the country's history, as the Cold War was beginning and McCarthyism was taking hold, the federal government was keen to embrace traditional American values as a way to demonstrate its stand against Communism.

Despite Brannan's major policy review and mounting concerns from members of Congress, the Department of Agriculture continued to push for increased modernization in agricultural practices. The concerns of Brannan and his colleagues never really diminished, and by the time Orville Freeman took over as secretary of agriculture in 1961, it was clear that something needed to be done.[6]

Freeman tried to push through a new vision of "an alternative agricultural landscape through new conservation programs that would support multiple uses of farmlands, including for outdoor recreation use, in an attempt to increase small farmers' incomes and preserve more diversified and beautiful agrarian landscapes."[7] Like his predecessors, Freeman praised the concept of the American family farm, viewing it as the ultimate symbol of democracy and progress. While Freeman was heaping praise on family farms, the darker and shadier side of the notion of progress began to rear its ugly head. The costs of technological advances were becoming altogether too apparent to ignore. Diminishing outdoor agricultural space, increasing poverty levels in rural communities and disappearing American heritage were deemed too high a price for society to pay, and so Freeman tried to create an alternative agricultural landscape to offset and coexist with the dominance of progressive farming in an effort to balance some of society's gathering tensions.[8]

Freeman's plans and proposals might have worked if it had not been for Earl Butz, the man who succeeded him as secretary of agriculture. During Freeman's tenure, farmers were required to conform to production control measures in order to decrease surplus if they wanted to receive price support payments. This helped to keep everything in check, especially since the country was still dealing with overproduction issues stemming from the increased production during World War II. Butz, however, made the executive decision that such methods of control were unnecessary and eliminated any and all production control measures. Many agricultural historians view Butz's reign as the tipping point in the world of farming—the point when the United States began its slide down the path of unsustainable farming practices. At the same time, reports of toxins and concerns over energy use began to emerge, as citizens and Joni Mitchell begged farmers to dispense with the use of pesticides because of their adverse environmental effects, loudly proclaiming, "Give me spots on my apples but leave me the birds and the bees."[9] Congress was forced to respond and moved to create a program for federal farmland preservation.[10]

Butz's intentions may have been well meaning, and Congress's response equally so, but not much was really being done to help the growing plight of family farmers. Families were losing their homes because they could not make their mortgage payments. For these families, the situation was rapidly approaching the same direness of the Depression. Though the government appeared to turn a blind eye, not everyone in America was quite as oblivious. In 1985, during his set at the Live Aid concert to raise funds to combat the famine in Ethiopia, Bob Dylan caused a ruckus when he asked whether or not any of the money being raised was going to be donated to the American farmers who were losing their homes.[11] Dylan's comments were not well received by some of his fellow performers. They did, however, catch the attention of Willie Nelson, John Mellencamp and Neil Young. Together, the three artists took a cue from the Live Aid concerts and organized Farm Aid, a concert to benefit American farmers. When asked why he helped organize the initial Farm Aid concert, Nelson was quite candid in his response: "I was just hoping to help farmers be able to farm food and feed their families… There was a time we were the strongest, after World War II, when everybody was pulling the plow. Now the government is only trying to make things better for the big corporations."[12] Since the initial concert in 1985, which was quickly followed by the establishment of Farm Aid as a nonprofit organization in 1987, more than $40 million has been raised and used to support family farms and to keep family farmers on their land.[13]

Farm Aid supports its mission to "keep family farmers on their land"[14] in four ways: promoting food from family farms, growing the good food movement, helping farmers thrive and taking action to change the system. To the first method, Farm Aid makes a point to feature food from family farms at its annual concert. The organization also works to bring farmers, artists and anyone with an interest in the plight of family farms together through television, mail and Internet campaigns. Farm Aid created homegrown.org as an online community where supporters of its cause can swap stories, connect and celebrate the American farmer. The organization also uses its resources to promote the good food movement. This involves making sure that family farmers have an outlet to sell their products and that consumers are aware of the products that are available. Next, Farm Aid is dedicated to helping farmers thrive. This has been the core of the mission since the first concert in 1985. It has created an online network for farmers to connect them to organizations around the United States so they can get the resources they need to flourish. Finally, Farm Aid works to change the status quo. By working with local, regional and national organizations, Farm Aid has fought to bring attention not only to the plight of family farmers but also to environmental issues arising from factory farming methods.[15]

Mellencamp, Nelson, Young and now Dave Matthews, who joined the Farm Aid board in 2001, are dedicated to this cause, even if they (Nelson in particular) wish the government would change its policies. "I wish I didn't have to do this," Nelson said. "I wish the government would take better care of our natural resources, and that includes the family farmer."[16] Mellencamp agreed, noting that Farm Aid will continue until it can no longer make an impact. "We all see what's happening with agriculture, what's happening to our small towns. They are going out of business. That's a direct result of the farm problem. We're still doing Farm Aid because it is contributing. It's doing a job."[17]

Part of that job is providing grants to both family farm and rural service organizations. In 2012, the grants went to programs that

> *help farm families stay on the land; build new market opportunities for farmers and increase consumer access to good food; confront corporate concentration in agriculture; advocate for fair farm policies on behalf of all family farmers, inform and organize farmers and eaters around issues such as factory farms, GE (genetically engineered) food, food safety and climate change; recruit and train new and beginning farmers and increase their access to farmland; and support farmer-to-farmer programs for more sustainable agricultural practices.[18]*

Each of the organizations receiving grant funds will help to alleviate some of the burden on family farmers, who are vital to the health of the United States. As an organization, Farm Aid supports the notion that independent farmers are the most important resource America has. Through its work with farmers, Farm Aid has seen that independent farmers "grow high quality food, are active in civic life, and are essential to the economic vitality of both their hometowns and the nation. As stewards of the land, family farmers work to protect the soil, air, water, and biodiversity in addition to producing high-quality, healthy food for everyone."[19]

In the years since Butz's mismanagement and the inaugural Farm Aid concert, the push for local and sustainable practices has really picked up steam, and while great strides have been made—GMO food labeling, the availability of certified organic products and more—the crusaders are fighting against more than 70 years of government investments in large farms and technological advances. Those 70 years might seem like quite a long time, but the 150 years preceding those advances showed that the concept of local and family-owned farms is prosperous and greatly benefits the community. Unfortunately, in spite of the increased awareness, a new demand for sustainable methods and less mass production, the federal government is still dragging its collective feet.

As 2012 drew to a close, Congress sat behind closed doors and deliberated on the 2012 Farm Bill, a direct descendant of President Roosevelt's Agricultural Adjustment Act. Unable to form a unified opinion, Congress finally agreed to extend the existing Farm Bill and to revisit the issue in 2013. In theory, the move was just fine. But in reality, the extension managed to do nothing except set farming issues on the back burner. Funds for programs supporting equity in farming and sustainable practices were not renewed. There were no changes to direct payment programs, despite a general consensus that cuts to these kinds of programs were acceptable and would be a great way to reduce federal spending. Probably worst of all, there was no mention of funding for disaster relief or to help farmers recover from a crippling drought.[20] In 2013, these issues will need to be addressed if the United States wants to continue to contribute, however minimally, to the success and livelihoods of small farmers. For his part, Willie Nelson will continue to fight for family farmers across the country because they are just too important to ignore. "I've always believed that the most important people on the planet are the ones who plant the seeds and care for the soil where they are grown."[21]

Nelson may firmly believe that the food producers are the most important aspect of strong farming community dedicated to sustainable practices, and he might just be correct. But there is one other huge factor to consider in the local food movement: food distribution.

Sustainable methods of agriculture are obviously important, but after taking the time and going through the trouble to use those methods, it is equally important to make sure that the products continue to have a minimal effect on the environment as they are transported to consumers. Herein lays the benefit of local food. Since many who embrace the locavore lifestyle choose to limit themselves to products grown or otherwise produced within one hundred miles of their location, they are helping to cut down on the environmental impact of food miles. The most common methods of distribution to fit in with local and sustainable values are direct-to-retail and direct-to-consumer markets. Direct-to-retail markets are steadily gaining a share in the local food movement. Usually this type of distribution scheme involves farmers selling and delivering directly to retail establishments. Some may choose to operate around a food hub, which the United States Department of Agriculture calls a "drop off point for multiple farmers and a pick up point for distribution firms and customers that want to buy source-verified local and regional food."[22] Food hubs serve a greater purpose in that they alleviate some of the pressures of marketing and transportation from the farmers. Food hubs can also serve to "expand the market reach of small, local farmers, help create local jobs, and can expand access to fresh, local food in urban and suburban markets."[23] Farmers also have the option to sell directly to their consumers. Usually these types of interactions take place through farmers' markets, CSAs and farm outreach programs, where consumers are invited directly to the farm to help harvest or pick their own vegetables.[24]

Great, but why does all of this matter? Farm Aid's work and the history of agricultural bills passed by the federal government have shown that there are severe economic consequences to family farmers if large-scale commercial farming endeavors get most of the business. There has also been some mention of the health benefits that come from supporting local farmers who use sustainable methods. All are valuable indicators of why local, sustainable eating is such an important part of life and why there has been such a great push to spread the word about this kind of lifestyle.

Economics first. The United States is just starting to pull itself out of its worst recession since the Great Depression era. Using local food systems stimulates local economies. It is a very simple cause-and-effect situation. Locally owned farms are often quick to give back to the community that supports them. And let's not forget about job creation. Small farms may not require as many employees, but they are more likely to hire those from their own communities when they do need someone. What it really boils down to is that local farms give back to local communities, and in turn, local communities give back to local farms.[25]

Then there are the environmental concerns, most often alleviated by using sustainable practices. Local farms tend to stave off making a negative impact on the environment by keeping pesticide use to a minimum, by treating the land with care and consideration through no-till methods and composting and by limiting food miles. While there has been no concrete evidence to suggest that food miles have a negative effect on the environment, it should be noted that local food only travels about 100 miles (according to the proscribed locavore style of living), while commercially produced food usually travels between 1,500 and 3,000 miles before it gets to the consumer.[26] Think of the carbon emissions coming from a delivery truck that only has to travel 100 miles versus one that travels 3,000. In the short term, it may not have much of a discernible impact, but in the long run, it likely will.

Finally, there are the nutritional and overall wellness benefits. Increased commercial production and global distribution practices have also increased the likelihood of food safety problems. The consolidation of processes provides more opportunities for bacteria to enter meat and produce. As food miles increase, the number of preservatives in food also increases. Finally, global centers of production tend to focus on the amount of output, rather than quality or nutritional value. Many times, these mass-produced efforts are then transported thousands of miles before they are purchased, resulting in less fresh products. Local farms, on the other hand, often have a fresher product because they focus on variety instead of quantity. The farmers take care to harvest produce at peak times, ensuring that quality. Plus, with fewer miles to travel from farm to table, freshness is virtually guaranteed.[27]

All of these factors feed back into the creation of a more well-developed society. Local farms support local economies. With local citizens buying local food, they are giving back to the farmers and local community. They are also increasing their own level of environmental consciousness and nutritional well-being. Making a minimal impact on the environment strengthens the planet. Increased nutrition makes for a healthier populace. A healthier citizenry helps lower healthcare costs. Such a seemingly small thing like supporting local and sustainable farms can have a much farther-reaching impact than it seems. True, there are many political factors to consider. If the information in this chapter has taught you nothing else, know that no matter how hard you try, it takes a long time for change to go through. However, through the efforts of communities supporting local and regional food systems around the country, and through the efforts of Farm Aid and its partner organizations, you can rest assured that as Sam Cooke once said, "A change is gonna come."[28]

Chapter 7

FARMING AND THE CITY

C ruise through the Rosewood section of Columbia on any given day, and you will see a sight that you may not expect to find in the middle of a city: a fully functional farm, complete with greenhouse. But Rosewood is where City Roots has, well, planted its roots.

City Roots is not only Columbia's but also South Carolina's first urban sustainable farm. Even with only about three acres of land, City Roots is able to produce 125 different types of fruits and vegetables. On top of that, the farm keeps bees, not only to produce honey but also to aid in the pollination process. It also keeps chickens for eggs and to help with fertilization of the soil. In order to keep the crop production so successful, City Roots uses methods of crop rotation and cover cropping to increase soil fertility and help with pest management. Further cementing the sustainable aspect of the farm, City Roots prides itself on its composting techniques, diverting items destined for the city's landfills into creating a fertile soil for the crops. Finally, using an aquaponic system—one that essentially mimics a typical pond and stream ecosystem—City Roots is able to raise tilapia and grow microgreens in water year round.[1]

Proud of its dedication to sustainable farming, City Roots has taken a holistic approach to farming, where each part plays a role in determining the success of other parts. Using all of the techniques and guidelines associated with organic farming, although to this point it has declined to undergo the inordinately expensive process of becoming certified organic, City Roots and its products have taken Columbia by storm. In not quite four years of existence, the farm has played host to five hundred volunteers, ten thousand event attendees, more than six thousand tour attendees and three hundred CSA members. It has also been honored with multiple awards, both locally and nationally, including the 2012 Green America's People and Planet Award, 2010 International Downtown Association Pinnacle Award, the 2010 Columbia Choice Award, the 2010/2011 Free Times Best of Columbia—Best New Green Business and the 2010 Farm City Award—Richland County.[2]

City Roots is the brainchild of Robbie McClam, who retired from the architectural field but still maintained an interest in farming. Thanks to a 2008 NPR segment on Will Allen's reception of the McArthur Foundation Award for his urban farm in Milwaukee, McClam started digging into the world of urban farming. Using Allen's organization, Growing Power, as a guide, McClam ultimately ended up registering for the Commercial Urban Agriculture Program in Milwaukee. Beginning in January 2009, McClam attended five monthly workshops in Wisconsin to learn all he could about urban farming. While in Wisconsin, the current Rosewood location was found, and McClam started the process for getting approval from the city council to open the farm. After working through a series of zoning issues and resolutions, City Roots was finally given permission to open as a working organic farm. Starting with only the basics—three acres, a used tobacco greenhouse and a few tools—City Roots and its staff have spent the last three and a half years hard at work.[3]

Of course, all of the hard work would not have paid off if there weren't at least some interest from the community. In fact, this much support has come as a bit of a pleasant surprise to McClam. "Our model, Growing Power in Milwaukee had many, many volunteers, interns and public recognition. Our fear was that we'd just be sitting down here next to Owens Field pulling weeds in total isolation. To our surprise, there was lots of interest and support right from the start."[4] It did not hurt that City Roots got so much publicity from the beginning. Between the zoning issues before city council and a front-page story in the *State*, Columbia's residents were well aware of City Roots almost before it started.

Building on an already established personal network that McClam had cultivated, City Roots emerged with a vibrant website and social media presence, spearheaded by McClam's son, Eric. Opportunities to set up shop at the All-Local Farmer's Market and the Healthy Carolina Farmer's Market at the University of South Carolina served not only to build a customer base but also as a way to recruit volunteers and interns. Since its inception,

Cellar on Greene's zucchini fritters are served with a tomato-basil coulis, pesto-arugula Israeli cous-cous, parmesan gremolata and everyone's favorite City Roots sprouts. *Photo by Clarissa Johnson.*

City Roots has fully embraced the local community and has thrown itself into participating in almost every local food and environmental organization it could find. "In essence our marketing effort initially was to just be very involved in the activities and organizations that we were passionate about," said the elder McClam.[5]

City Roots was also able to spread the word about its produce by partnering with several restaurants in Columbia. Though this had been a goal from the very beginning, the McClams were restrained in their approach. "We were cautious at first about marketing to restaurants because we didn't want to begin that effort until we felt confident that we could meet the demands of the chefs on a consistent basis," Robbie McClam said. "I still remember Eric and I personally visiting several of the restaurants and having to explain just what microgreens were and how nutritious and flavorful they were. Once the chefs experienced them and believed they could depend on us for reliable supplies the produce sold itself!"[6]

Building on existing relationships with local stores and restaurants, the McClams were able to partner with establishments like Rosewood Market,

Gervais & Vine, Rosso, Mr. Friendly's, Cellar on Greene and Solstice. Working with Basil Garzia at Rosewood Market was the first step in an ongoing collaboration process. Before the first planting at City Roots, a plan was already in place to pick up Rosewood Market's produce waste and use it in the farm's composting program. In return, Rosewood Market became the first place in town to sell City Roots microgreens. From there, conversations with Kristian Niemi, then owner of Gervais & Vine and Rosso, and Ricky Mollohan of Mr. Friendly's, Cellar on Greene and Solstice fame launched City Roots microgreens into the Columbia restaurant scene. Now, City Roots has its produce featured at about twenty restaurants and sold in six grocery and health food stores.

With so much success in the Columbia area, expansion to other regions is definitely in the cards. In fact, it is already happening. Beginning in the early part of 2012, City Roots started marketing to various restaurants. While there was certainly a large interest in the produce, City Roots' small staff found themselves spending more time driving to deliver the produce and building relationships than they were equipped for, so unfortunately, the enterprise was temporarily shut down. Later in the year, City Roots tried again, this time aligning itself with the Charleston-based nonprofit Grow Food SC. Robbie McClam is pretty excited about the partnership. "Their mission is to work with local SC farmers like us to get our produce into the kitchens of the restaurants and on the shelves of the food retailers. Our mission is to 'produce clean, healthy, sustainably grown products,' so for us it was a perfect match—we grow it and they sell it!"[7] As a result of this newly forged partnership, City Roots has taken the Charleston area market by storm, with more than seventy-five restaurants and retailers using its products.

So, now that the farm has established itself in Columbia and has a strong foothold in Charleston, what happens next? There are some definite plans, even if there is not much free time to think about them. As the staff gets better at both the business and profession of farming, an occurrence that McClam swears would never have happened without the talents of Ben Dubard, they are looking forward to expanding their acreage.

Perhaps the best thing to come out of all of City Roots' overnight success is that the farm received a Conservation Innovation Grant from the U.S. Department of Agriculture. As a result of the grant, the farm is implementing a no-till farming process that will basically serve to make its daily operations more energy efficient and more environmentally friendly. Increased efficiency is one of the desired results of this grant. If it proves effective, City Roots will be able to generate more income and bring on more staff, which in turn can only mean good things for the expansion of the farm. It may even lead to expansion into the Lowcountry. Charleston is a virtually untapped market for urban farming, and City Roots is seriously contemplating opening City Roots part two in the area.[8]

City Roots has clearly surpassed everyone's expectations. The farm's popularity has taken the McClams by surprise, but they have lovingly welcomed the community's embrace. Said Robbie McClam:

> *For some reason, our farm has resonated with a lot of people and for that we're very surprised and appreciative. The personal stories have been incredible, from Geordan and Gee, the 10 and 12 year old sister and brother that adopted us one summer and made us their home, to the unemployed but energetic who just want something meaningful to do, to the thousands of school kids that have squealed and laughed at the chickens and tilapia. One only has to watch middle schoolers fight to spend their own money on your purple carrots and yard long beans to know you might be onto something special. Columbia is indeed ready to be a part of the local, sustainable food movement and we're proud to have a role in that development.[9]*

Chapter 8

VEGGIES AND HERBS AND PUMPKINS, OH MY!

It would be impossible to write a book about local, sustainable food without at least mentioning the importance of the CSA. Community supported agriculture (CSA) has become a huge deal in the food world over the course of the last twenty years. Basically, CSAs are a way for consumers to buy directly from local farmers. The farmers offer a share to the public, normally in the form of a box of seasonal vegetables. Consumers pay a membership fee as a means of purchasing a share and then receive a box directly from the farm once a week for the duration of the season. It is a simple-enough process, with tremendous benefits to both the farmers and the consumers. To start, it allows patrons to eat seasonally (more healthily) and to discover new uses for food. For the farmers, CSAs are simply a good business model that allows them to take care of marketing concerns early so they can focus more attention on the growing and harvesting of the crops.[1] The best part is that CSAs are seasonal. If winter veggies are not really your thing, you can simply skip the winter season and pick back up in the spring.

In Columbia, folks interested in a CSA have five options: Round River Farms, City Roots, D&J Farm, Doko Farm and Pinckney's Produce. Each farm offers different size and price options, so there is something to fit every budget. Just think of it: a weekly delivery of fresh produce that will keep you from wasting your time scouring grocery store chains in search of good-looking yet reasonably priced veggies.

Though it is not technically located in, or even around, Columbia, Pinckney's Produce, headquartered in Beaufort, has been providing CSAs to the Columbia area since 2009. Like most CSAs, Pinckney's operates by having members prepay for a share of the season's harvest. Then, when the harvest comes in, it delivers produce boxes to a local drop site once per week. Pinckney's has something for everyone, offering four different share sizes in each season. Both the fall and spring seasons are twelve weeks longs, meaning that each shareholder receives twelve boxes of fresh veggies from April through June and again from September through December.

Pinckney's delivery-based program has been the catalyst to expansion into larger markets outside the farm's immediate area. The support the program has received in Columbia has even allowed Pinckney's to deliver twice a week during the season. With more than fifteen drop-off sites in town, there is a spot for everyone, from Lexington to Blythewood and everywhere in between. Pinckney's credits the relationships with local businesses for their success in Columbia. As Katie Thompson, membership coordinator, noted, "Our program thrives on a strong relationship between our members, our drop sites and local business owners and the farm. We think mutual support is key to creating and building a strong local food community."[2]

The community is very strong indeed—so strong, in fact, that it prompted owners and creators Pinckney Thompson and his father, Dr. Jody Thompson, to move in December 2011 from the Thompson Farm in Holly Hill, South Carolina, to Rest Park Farm in Beaufort, where father and son team Urbie and Ashby West took over responsibility for CSA production. Rest Park Farm is more than eight hundred acres, and the coastal climate has allowed for a stronger and longer growing season, meaning that Pinckney's can provide a full fall season worth of vegetables in addition to its immensely popular spring shares.

What can you expect to find in a CSA from Pinckney's Produce? Good question. Obviously, it will depend on the season. But in each fall and for each spring, more than forty-five different varieties of produce are offered to members. You will find typical favorites like squash, sweet corn, tomatoes, melons and broccoli, in addition to more unique items like tat soi, daikon and bok choi. And because you will need to season your vegetables as well, Pinckney's also includes specialty items like herbs and hot peppers. It also partners with other local specialty farms to include blueberries, strawberries or even pumpkins in its boxes.[3]

Columbia blogger April Blake is well known for her vegetarian food blogging and for her focus on healthy eating. In an effort to combine those interests and to challenge herself to eat more seasonally and locally in the spring of 2012, Blake did her research and decided that a CSA was the best way to meet the challenge head on. After extensive research, she eventually decided that Pinckney's Produce was the way to go. The flexible delivery schedule and locations were a large part of the decision to choose Pinckney's. "The drop off or pick up locations and times for other CSAs were just too far out of my way, or the shares had to be picked up by 5 p.m.—something that I couldn't do with my work schedule."[4]

Schedule and location weren't the only reasons why Blake has fallen in love with the options from Pinckney's. The huge variety of fruits and vegetables provided also piqued her interest:

> *The variety of vegetables and fruits in the boxes are amazing, though in the cooler months of the year sometimes I would be like, "Ugh, not more collards!" In the summer I also had similar moments with eggplant after a few weeks of the enormous purple bulbs still showing up in the box. But that's part of eating seasonally, and I used the swap box that they provide at drop off locations to trade out my hated collards for turnips, cabbage, or other greens, or I'd trade eggplants in for more easily preserved items, like tomatoes, peppers or corn.*[5]

After two seasons with a Pinckney's CSA box, Blake has learned a few tricks. She now tries to preserve what she can't use immediately and freeze what she could not eat. Thanks to her planning, she can eat fresh summer corn, frozen mere days after it was picked, in January. Blake has also found herself trying more and more new foods as a direct result of the CSA. She has discovered new loves in tat soi, cauliflower and cabbage, vegetables that she swears she never would have tried if it hadn't been for her CSA. "I look forward to my new box full each week and find that I can put off grocery store trips more because I focus on eating the vegetables more before they go bad or I get another haul of vegetables the next week!"[6]

It seems that Pinckney's Produce has found a fangirl in April Blake. She had nothing but raves for the CSA program and the people who run it. "The folks at Pinckney's are fantastic and have great pride in the South Carolina grown produce that they harvest."

With support from people like Blake, Pinckney's should have no trouble meeting the goals it has established for the future. "We want to continue to nurture our food community and watch it continue to grow. We hope to benefit many people's lives, our local communities and agriculture for years to come," said Katie Thompson.[7]

Chapter 9

HAPPILY EVER AFTER IN THE MARKETPLACE

Close your eyes and imagine a brisk, autumn Saturday in Columbia. There is a slight bite in the morning air, but there is the definite promise of a sunny and warm afternoon. Or maybe it is a hot and humid morning in late spring or early summer, the kind of day where leaving the house in the afternoon is not an option unless submerging yourself in Lake Murray is the end result. Either way, the perfect start to this kind of day is a trip to a farmers' market to stock up on produce and meats to grill for a football tailgate or prepare that giant salad that won't heat up the kitchen in the summer.

Around the country, on days like these, people everywhere are stocking up on local produce to support their respective communities. In Columbia, folks have started flocking to Soda City market. Featuring meats, produce and crafts from Midlands-area vendors, Soda City is the brainchild of Columbia farmer Emile DeFelice. An open-air market taking up an entire block of Main Street each Saturday morning, Soda City features a constantly changing lineup of vendors, designed to showcase the best the Midlands has to offer.

Soda City, as it is now known, started in 2005 as the slightly more humble All-Local Farmer's Market. Bouncing from location to location, the market found a more permanent home in the space next to 701 Whaley in 2009. Creating an environment reminiscent of a Middle Eastern bazaar, patrons rubbed elbows with local vendors, nourished themselves with food from trucks or featured local restaurants and often enjoyed music from local acts.

Word of the market and its goodies filtered through Columbia at a rapid pace, and within three years, the All-Local Farmer's Market was bursting at the seams and desperately seeking a new home. After eight years, it was time to transform the market model into something more sustainable. For DeFelice, that meant a "name, location, content, and goal shift."[1] In October 2012, the first shot at transformation took place with the opening of Soda City.

After the relatively constraining space restrictions in the space on Whaley Street, the new market took over the entire 1500 block of Main Street and used it to its advantage. Soda City added new types of vendors, focusing on more than just the local food aspect of Columbia. Soda City embraced all things local, featuring artists, artisans, retailers, collectors and even a lawn and garden area. And, of course, the focus on South Carolina farmers—whether the product was produce, meat or flowers—remained.

In just three short months, Soda City became what owner and operator DeFelice called an "unqualified success."[2] Given the popularity of the previous market, the new location, which also allows patrons a chance to stroll along Main Street, it is not surprising that Soda City has elicited such a positive response. Soda City has also benefited from extraordinarily positive responses from the vendors as well. In light of Mayor Steve Benjamin's ongoing quest to revitalize Main Street, Soda City was a major boon to the city. In fact, when it opened, Soda City was the first, full-fledged farmers' market to cater to Columbia's citizens in over fifty years.[3]

Though the lineup of vendors changes weekly, Columbia-area carnivores have had the option to browse offerings from Caw Caw Creek Farms in St. Matthews, Sea Eagle Seafood in Beaufort and Swansea Farms and Will-Moore Farms in Lugoff. Omnivores get the best of both worlds, with exposure to all of the meat and dairy options in addition to products from City Roots Farm, Anson Mills, Crooked Cedar Farm and Gnome Grown. And that's just what is for sale to take home. Soda City also features several local eateries that provide their own brand of brunch for hungry shoppers, including area favorites Rosso, Crepes & Croissants and Pawley's Food Truck. To top it off, Soda City even caters to the crafty crowd, with goods from Backcountry Skin Care, Bill's Birdhouses and Sixteen Acre Wood.[4]

As befits a local market, the list of vendors is constantly changing. About half of the vendors on any given Saturday remain the same, but the other half rotate in and out. An even more diverse assortment of goods is expected as the seasons change. But is that not the best part of a farmers' market? Like Forrest Gump and his box of chocolates, you never know what you're going to get.

The response to DeFelice's pet project from vendors and residents alike has been "terrific, supportive and excited and helpful in every way."[5] Soda City has even prompted locals to endorse it through more than just word of mouth. Longtime Columbia resident Tom Prioreschi submitted a letter to the *Free Times*, detailing the reasons why Soda City is a wonderful thing for Columbia. Normally, letters like this only crop up when a landmark is in danger of shutting down or being destroyed. In this case, however, Soda City is just getting started; Prioreschi's letter appeared a mere two months after the market took over Main Street. The letter described the benefits of Soda City's residency in terms of economic impact, quality of life and general excellence. As he wrote, "This market will push Main Street past the tipping point of sustained energy and will change the future dynamics of Main Street."[6] DeFelice has high hopes of influencing those same changing dynamics. Though Soda City has just begun its run as a Columbia mainstay, DeFelice already has plans to incorporate a brick-and-mortar store and other markets to further enhance the experience for everyone involved.

These future plans should make local food blogger Bach Pham happy. Even though he was skeptical about Soda City at first, he has come to embrace it and everything the market represents:

The previous market at Whaley's had such a great local vibe to it that seemed hard to beat, especially since it was well on its way to becoming an established fixture of the Olympia neighborhood at that point. When Soda City arrived, however, I was blown away immediately. The open-air atmosphere and charisma of the vendors made it a real treat to attend. Quite suddenly, alongside Mast General Store, the opening of the Nickelodeon and Paradise Ice, Main Street has become the place to be, which for anyone who has lived here for a while knows that that's not a praise you hear often in the Capital City. The stores in particular have really elevated Soda City's status, adding more flavor and variety to the market than the former could have achieved. And we can't forget the people: between the original vendors like City Roots and their incredible organic vegetables, Caw Caw Creek and their amazing local pork, and Rosso's spectacular brunch that made Whaley's All Local Farmer's Market so great, and the new vendors like Crepes and Croissants that have contributed to Soda City's stability and success, everyone has worked very hard in making what could have been a great failure an enormous success.[7]

Of course, as with every new endeavor, there are those who prefer the status quo. In this case, Frank Adams is dedicated to keeping the spirit of the All-Local Market at 711 Whaley alive and thriving. While he certainly respects the efforts and the all-encompassing nature of Soda City, Adams prefers the vibe of the original market. In his view, more markets help support the notion that everyone should be united in the effort to promote sustainable agriculture and unadulterated foods. According to Adams, "We need dozens of markets that bring forth the best of what people make and grow, and we need to support the concept of enjoying life with good friends, acoustic music, simple foods and a comfortable interchange of ideas and mutual goals. That is what a community is all about. It's what I look forward to each Saturday morning."[8]

A vast majority of Columbians would agree. The ongoing sense of community, and the pride taken in labor and craftsmanship of those in the Capital City, is what will keep both the smaller, established markets and the bigger, "new kid on the block" Soda City in business and on the path to success.

Chapter 10

THIS LITTLE PIGGY WENT TO MARKET

When the Soda City market hit the Columbia food scene in October 2012, there were some locals who were not too sure about it. While everyone can agree that a larger area in which to showcase local products, artisans and farmers is a good thing, some felt that the new space lacked some of the true community aspects that had made the previous All-Local Farmers' Market so enjoyable. And so, out of the ashes of the previous market arose a new venture, one that really put the emphasis on building the community spirit that threatened to dissolve as Soda City began to dominate the Saturday morning market landscape. Enter the Vista Marketplace at Whaley.

In August 2012, with the cloud of the impending fate of the All-Local Farmers' Market hanging over the space on Whaley Street, the Vista Marketplace opened its virtual doors. Yes, that is correct. Vista Marketplace started (and continues) as an online vendor. It is a community of local craftsmen, artisans, farmers and other local business owners who sell their products through both the traditional booths at craft fairs and farmers' markets but who also appeal to the state's twenty-first-century sensibilities by offering their products through its website. Through a CSA-type program, buyers are able to place an order for exactly what they want, in terms of quantity and type of product, and about one week later, those products are delivered to a central location where buyers can pick up their goods.[1]

Once the space at 711 Whaley Street started to resemble a ghost town on Saturday mornings, the Vista Marketplace added another dimension to its already established online distribution by setting up a smaller-scale market on Whaley Street. With a limited number of vendors, who change based on availability, the focus at Vista Marketplace is really on creating a sense of community.

Frank Adams is a longtime Columbia resident who is always searching for that community. He thinks he has found it on Saturday mornings at the Vista Marketplace. "To me, people who are artistic, who believe in whole and unadulterated foods, who have made their own beer, who walk into their kitchen hungry, but with no plan but to create, people who want to share and be friendly, people who understand the concept of savoring—these people are a community. I find them wherever I find originality and creativity, and I welcome their company."[2] With the gathering of the minds that the Vista Marketplace has brought together, it is not surprising that Adams is so supportive.

Vista Marketplace makes a point to include local musicians and feature brunch from local restaurants. It is open for four hours late on Saturday mornings through early Saturday afternoons. Due to the nature of the space on Whaley Street, even if you finish shopping early, the temptation to stay and relax at one of the picnic tables while listening to local musicians is a strong one. The Vista Marketplace at Whaley has yet to go through a spring or summer or even early fall season yet, but once the weather turns a bit nicer, all of those creative types that Adams alluded to should come out of the woodwork to spend Saturdays with other like-minded folks at the Vista Marketplace.

Fresh, open air–style farmers' markets are wonderful additions to any community, but what happens when devoting an entire Saturday morning to browsing the stalls just is not in the cards? At that point, do you just resign yourself to hitting the closest grocery store and making do with its produce section? You could. Or you could invest in a CSA. While most CSAs do have a set pick-up time, options are usually offered during the week that provide the perfect excuse to get out of the office for lunch. Grab something to eat and pick up some fresh veggies. Done and done.

There are a few options for CSA membership in the Columbia area. There are different sizes, pick-up dates and costs, designed to suit as many people as possible. There are two big CSA providers—City Roots and Pinckney's Produce—that are discussed in separate chapters of this book. Do not despair, however, if you are of the "Damn the man! Save the empire!"[3] mindset and want to choose a smaller farm for your CSA (not that City Roots and Pinckney's are anything like fictional corporate giant Music

Town).[4] There are still three great options to choose from in the Columbia area, with several more choices throughout South Carolina.

Tucked away in the northeast section of Columbia, splitting the distance between I-77 and I-20, is D&J Farm. A member of the Southern Sustainable Agriculture Working Group (SSAWG), D&J Farm specializes in naturally grown products. What that means is that the farm has not chosen to become certified organic but still follows the principles and practices of organic farming.[5] D&J Farm only offers CSA shares in one season, from June through October, for a full twenty-one weeks. The farm's produce options range from beets to potatoes to peanuts to zucchini. Members choose whether to receive a half share (seven pounds of produce per week) or a full share (twelve pounds of produce per week). As an added benefit, the farm will add a watermelon to each box twice a month. Maybe the best part of D&J Farm's CSA package is its delivery methods. CSA boxes can be delivered to the Columbia area during the day on Fridays, either to the member's home or office.[6]

About thirty miles to the southwest of Columbia lies Round River Farms, another option for anyone interested in investing in a CSA share. The farm is twenty-two acres and features a pine and hardwood forest, an organic garden, pastures and a certified wildlife habitat.[7] Using the square foot gardening method, Round River Farms produces assorted herbs and vegetables from May to October. Each CSA share will contain produce including broccoli, bell peppers and edamame, as well as various herbs, depending on the season. The farm designates pick-up sites in Lexington and Columbia and delivers twice per week. The folks at Round River Farms are extraordinarily invested in their CSAs. On their website, they firmly define their commitment to the general principles of CSAs. "Community Supported Agriculture allows consumers to take the responsibility for the food system which nourishes them. It allows social responsibility toward the land. Agricultural workers need to be respected and supported and the land needs to be respected and cared for. When this is the case, the food which is produced will be truly nourishing."[8]

Doko Farm is truly a family affair. The farm sits on land in Blythewood, South Carolina, that has been owned by Joe Jones's family since the mid-1800s. Since then, at least one family member in each of six generations has lived on the farm.[9] With family ties to the land and a deeply embedded commitment to sustainability, the Joneses are intent on providing people with the chance to really develop a connection to what they eat. In a 2011 interview with the University of South Carolina (Joe Jones received two

Prepping mushrooms for dinner at Terra. *Photo provided by Terra.*

Dinner by Terra. *Photo provided by Terra.*

Fresh fish at Terra. *Photo provided by Terra.*

Walking the fields at City Roots. *Photo by Robbie McClam.*

City Roots hosts a Harvest Dinner. *Photo by Robbie McClam.*

The Hardee family, relatives of current farm owner Urbie West, on Rest Park Farm, which is now home to Pinckney's Produce. *Photo by Claude McLeod.*

Above: Tilling the fields on Rest Park Farm. *Photo by Urbie West.*

Left: Working the fields on Rest Park Farm. *Photo by Urbie West.*

If you look closely, you can see the way each ingredient melds into the zucchini fritters. *Photo by Clarissa Johnson.*

Lowcountry collard greens serve as an accompaniment to pork tenderloin during dinner at Mr. Friendly's. *Photo by Clarissa Johnson.*

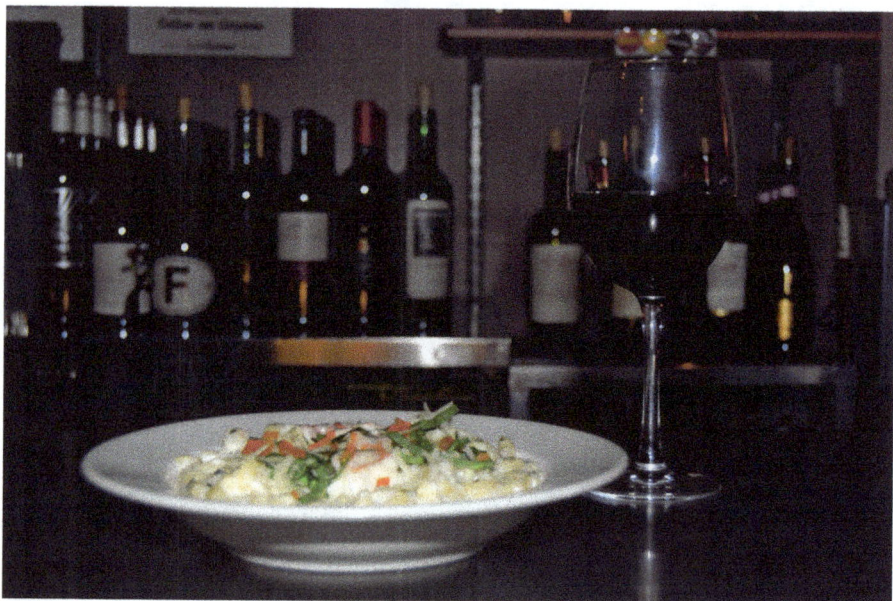

Lobster mac and cheese with a glass of wine for an evening at Cellar on Greene. *Photo by Clarissa Johnson.*

Creamy "French Quarter" pimento cheese is the perfect topping for Mr. Friendly's famous bacon-wrapped angus filet mignon with roasted garlic mashed potatoes and sautéed seasonal veggies. *Photo by Clarissa Johnson.*

Toasted nuts for sale at Rosewood Market. *Photo by Clarissa Johnson.*

Fresh baked scones at Rosewood Market. *Photo by Clarissa Johnson.*

One of the many baked goods for sale and consumption at Rosewood Market. *Photo by Clarissa Johnson.*

A quick look at Rosewood Market's produce section. *Photo by Clarissa Johnson.*

A selection of locally grown and organic produce at Rosewood Market. *Photo by Clarissa Johnson.*

Tomatoes grown just twenty minutes from Rosewood Market. *Photo by Clarissa Johnson.*

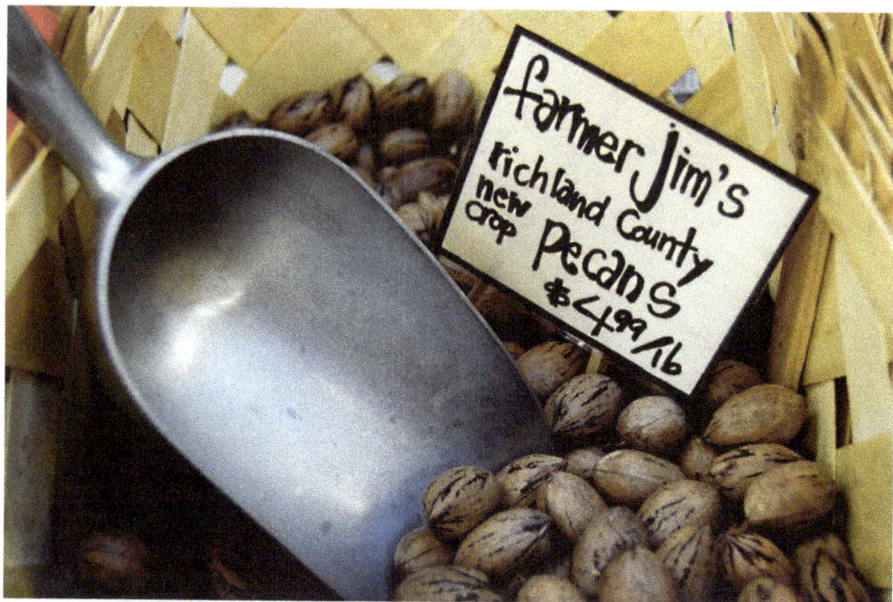

Making a pecan pie? Stop in at Rosewood Market to pick up some home-grown pecans. *Photo by Clarissa Johnson.*

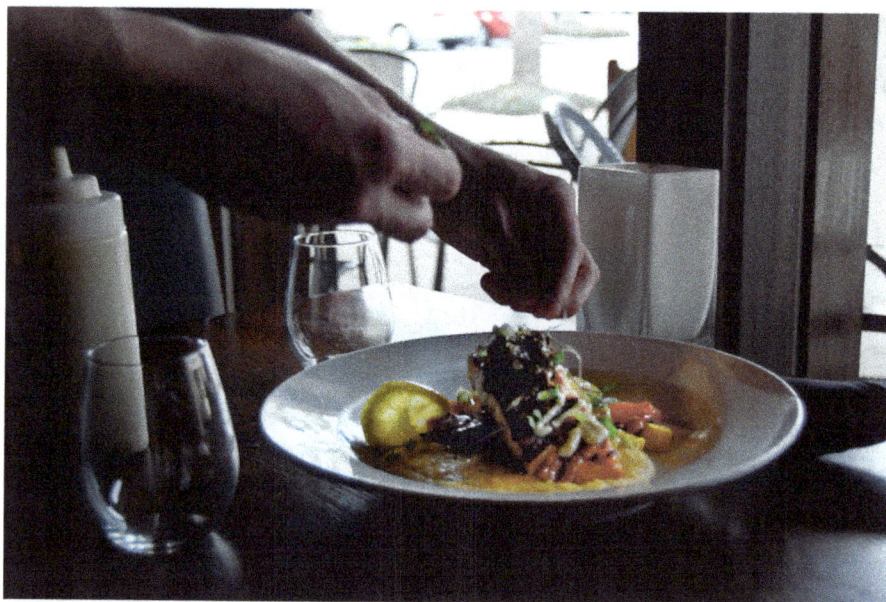

Adding just a hint of City Roots microgreens to the salmon at Rosso. *Photo by Clarissa Johnson.*

The salmon is so tender that there's no need for a knife. *Photo by Clarissa Johnson.*

Staring at the Columbia skyline for dinner at Terra. *Photo by Clarissa Johnson.*

Just think about how that creamy goat cheese will blend with the sherry vinaigrette dressing. *Photo by Clarissa Johnson.*

Crispy frogs are what's for dinner at Terra. *Photo by Clarissa Johnson.*

Chef Mike Davis prepares for dinner at Terra. *Photo provided by Terra.*

Set up for a Harvest Dinner inside the greenhouse at City Roots. *Photo by Robbie McClam.*

Giving a tour at City Roots. *Photo by Robbie McClam.*

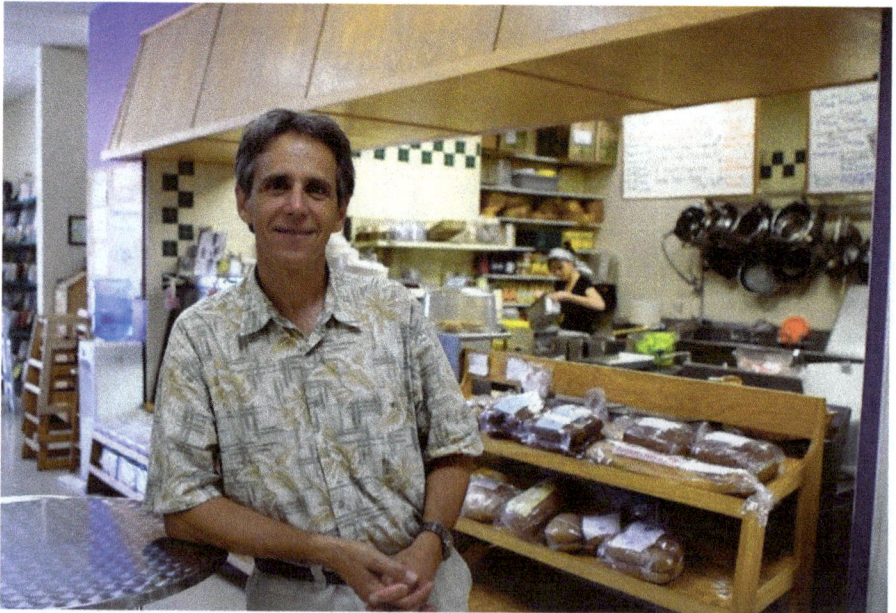

Rosewood Market owner Basil Garzia surveys his domain. *Courtesy of Rosewood Market.*

Above: Motor Supply Company's Chef Tim Peters and owner Eddie Wales. *Photo by Jonathan Sharpe.*

Left: Shrimp and grits at Motor Supply Company. *Courtesy of Motor Supply Company.*

Farm-fresh eggs from Wil-Moore Farm. *Photo by Keith Willoughby.*

Keith Willoughby feeds the chickens at Wil-Moore Farm. *Photo by Zac Willoughby.*

degrees from the university), Amanda Jones explained why it is so important to make that connection: "There's a deeper meaning to food when you raise it yourself or buy it from a farm where you can see for yourself how things are raised…People like to come to our little farm just to see how we raise the livestock and to experience a connection with what they're eating. You can't do that in a grocery store or a restaurant."[10]

Not only does Doko Farm provide consumers with the opportunity to foster a relationship with what they are eating, it also runs the only CSA with a focus on meat in Richland County. You read that correctly. Doko Farm's primary CSA share is devoted to heritage meats. Some things you could expect to receive in a CSA share from Doko Farm might be guinea hog pork, Saxony duck or the slightly more recognizable lamb. And, of course, once November rolls around, customers have the option to receive a Narragansett heritage turkey, perfect for feeding the family on Thanksgiving.[11]

Chapter 11

TO MARKET, TO MARKET

How do you handle this situation? The farmers' market is closed, and you forgot to sign up for your CSA. You want something fresh, nutritious and local for lunch or for dinner. Where can you go? The answer is simple: Rosewood Market. Rosewood Market sells products from many of the same vendors who set up shop at Columbia's various farmers' markets on Saturdays, and then some.

Around the same time that *The Godfather* was winning the Academy Award for Best Picture, Basil Garzia was opening the Basil Pot, a vegetarian joint on Rosewood Drive. The Basil Pot finally closed its doors in 2003, after a change in ownership and a move to Main Street, but the philosophies of the restaurant still remained. The main idea that has carried over into the fabric and foundation of Rosewood Market was that "people can take an active, hands-on approach to their own wellness through delicious food."[1] It is an idea that had a quite a bit of merit, since now, about forty years later, the market is still thriving.

Part of Rosewood Market's longevity can be attributed to its strong ties to the community. Local vendors, craftsman and community members, not to mention the staff and customers, have shaped Rosewood Market's success. The market's continued commitment to all things local, environmental and health-related has really provided Columbia's residents with the services of some larger chain health food stores but in a more personable and almost intimate way. That relationship building has led to multiple generations of families investing their time and money into keeping Rosewood Market up

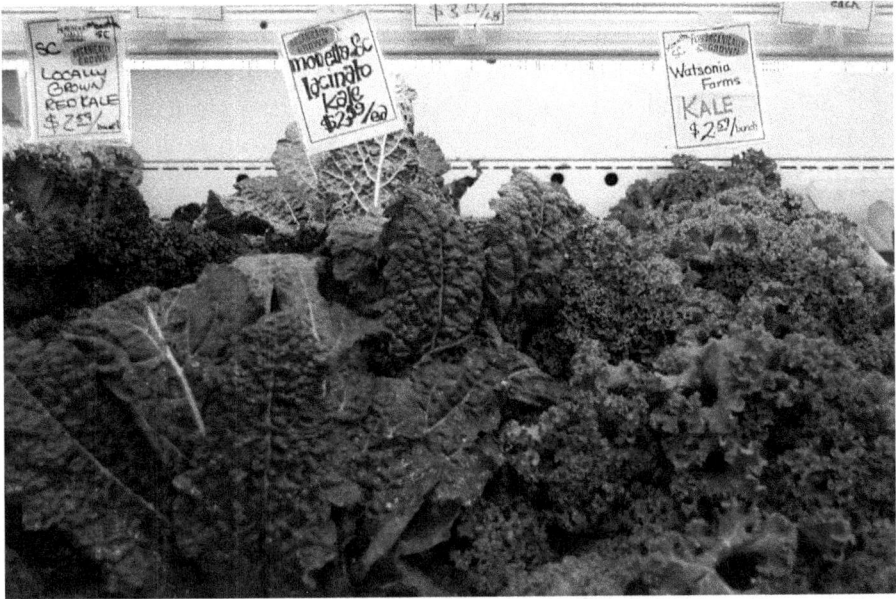

South Carolina–grown organic kale at Rosewood Market. *Photo by Clarissa Johnson.*

and running. "The success of Rosewood Market derives from the consistent, steady work and devotion on the part of its staff members, several of whom have been here twenty years or more," Garzia explained.[2]

As expected with any type of market, there are many things that would traditionally be seen as waste. Rosewood Market, however, because of its commitment to environmental health, has tried to minimize its waste. By working with several area partners, and in some cases just hoping a random stranger helps them out, Rosewood Market has been able to recycle 100 percent of its cardboard, magazines, newspapers and office paper. Any kind of compostable waste from the kitchen or the produce department is taken by City Roots, which then takes care of the composting products. Empty buckets and crates are left on the curb as on offering to anyone passing by and are only disposed of on the rare occasion that nobody claims them. Even items that do not yet have recyclable outlets are not thrown away. Batteries, fluorescent tube lights, phone books and even wooden pallets are stored at Rosewood Market until they find someone to properly recycle them. And just to be prepared for when someone in Columbia starts to compost nonorganic matter, Rosewood Market's deli serves all of its food on compostable plates with compostable utensils.[3]

The environmental efforts are commendable, but at its heart, Rosewood Market is just that—a market. As you would expect from an organization so deeply embedded in the local culture, it features products from local vendors, as well as typical natural food store fare, from bulk products to health and wellness products. Sometimes products are introduced by employees and customers. Orangeland Fish & Seafood in Florence, South Carolina, was brought on to the list of Rosewood Market's vendors when an employee made the suggestion. Orangeland Fish brings in whole fish, blue crab and never frozen, always fresh shrimp that have been caught the day before to serve Rosewood Market's customers. Similarly, Rosewood Market was introduced to Farmer Jim by a longtime customer. Farmer Jim supplies the market with blackberries, cucumbers, pecans, okra, tomatoes and, in the autumn, muscadine and scuppernong grapes. Sometimes Farmer Jim delivers his contributions directly to Rosewood Market's produce manager, and other times he sets up shop in the market, and customers can interact and buy directly from him.[4]

As with most successful businesses, Rosewood Market is very keen to meet the needs of its customers. With so many regular patrons, the staff has taken suggestions and made them happen. As veganism gained popularity, Rosewood Market upped its supply of vegan-friendly products. To cater to the needs of those with gluten issues, space on shelves and in refrigerators was made to stock gluten-free products. Even Rosewood Market's deli, which has an enormous offering of prepared and made-to-order foods, has adjusted its menu to meet the needs of the customers. Now, entrées and sandwiches can be made gluten free. Deli items are also clearly marked to show whether they fit macrobiotic, vegetarian or other standards.[5]

Part of Rosewood Market's enduring appeal is the friendliness and knowledge of the staff. Staff members are quick and eager to explain the meaning of things on food labels, discuss ingredients and make suggestions about products. One customer interviewed in Joan Hardy Eison's article mentioned the staff as one of the best things about Rosewood Market. "Very friendly, helpful, knowledgeable staff who care about what they do."[6] Part of that may even include knowing the names of customers as they walk through the door. But that has been Garzia's intention all along. "I like to say we are in the people business, providing groceries and meals on the side. We like what we do and what we offer our customers. As a result, it's become a mutual admiration society," Garzia said.[7] In short, loyalty is built by customers who feel welcome and respected every time they walk through Rosewood Market's door. As another customer in Eison's article stated, "I love the hometown feel and pleasant staff. It feels like wandering into a home pantry."[8] And that's the point of a market that puts its customers' needs above its own and specializes in supporting local vendors and the local environment.

Chapter 12

THE SALT OF THE EARTH

Whether you are looking for upscale dining or a more casual evening out, as long as you are searching for a fresh, simple and locally focused menu, Terra is the place to be. Featuring a menu that changes not only seasonally but also daily, Terra chef Mike Davis makes it a point to feature as many fresh, local ingredients as possible in his daily creations.

Chef Davis brought his culinary expertise to town when he opened Terra in West Columbia in 2006. But it was long journey to get there. Raised in a farming family in Alabama, Davis found his way into the kitchen during his undergraduate years at the University of Alabama. In an effort to earn money to spend on his extracurricular pursuits, Davis took a job at an Italian joint called DePalma's that specialized in pizza and pasta. While initially just a way to earn some extra cash, his stint at DePalma's set the stage for Davis's future forays in the kitchen.

After graduation, Davis was really bitten by the cooking bug, and so he enrolled at Johnson and Wales in Charleston. Since enrolling, he has worked in some of the finest restaurants in the country—Magnolia's in Charleston; Bayona and Cobalt, both in New Orleans; and Chez Fon Fon in Birmingham—under the tutelage of several well-known chefs. In particular, James Beard Award–winning chefs Susan Spicer of Bayona and Frank Stitt of Chez Fon Fon influenced Davis's future pursuits. It was actually during his time at Chez Fon Fon that fate struck, and a restaurant went up for sale in Columbia. With ties to the Columbia community—his wife is a native Columbian—Davis purchased the space and Terra was born.

Terra's intense focus on all things local can be traced back to the time Davis spent working for Chefs Spicer and Stitt. Chef Stitt's philosophies seem to have been the most influential for Davis:

> *His* [Chef Stitt's] *philosophy on food was the reason I went to work for him. He basically took French sensibilities towards food and applied them to Southern food. This fascinated me, and I have been a believer since then. He worked for Alice Waters in California, who pretty much started the whole farm to table thing as we know it now. It birthed California cuisine, and he applied those principals to food in Alabama. Basically, cooking with the season and using what is coming from the earth at that time of year to base your menu on. Food is obviously freshest if you buy it from the person that grows/raises it. The fresher, the better, more beautiful and so on. It made sense to me; a tomato in July is far superior to one in February, it grows naturally in the summer, hence it should be eaten in the summer. As I dug deeper into cuisine, I realized that this was smart and tasty, and this would be what I based Terra on.*[1]

And so that is what Chef Davis has done. He has taken traditional cuisine and added a Columbia flair so that he is always focusing the freshest produce, fish and meat that essentially come from Terra's own backyard.

When Chef Davis and his vision for Terra arrived in Columbia in 2006, the city was in dire need of a "good restaurant and one that took advantage of all the wonderful foodstuffs that are produced around here."[2] As such, Terra filled the void, broaching virtually uncharted territory in the Columbia restaurant scene, and it has been going strong ever since. "Six and a half years ago, not many restaurants thought like this, but it has now become common in the restaurant world. It has because it makes sense. As far as what will happen in the future…I am not sure, but eating healthy, knowing where your food comes from, and eating the freshest possible food will hopefully help the American people to better their diet and our country."[3]

With such an intense focus on using local ingredients, you might think it hard for Terra to vary its menu too much. Thankfully, that is not the case. Between Chef Davis's culinary creativity and the plethora of local ingredients, there is always something new going on in Terra's kitchen. "We work seasonally by serving what is coming from the land in that season…There are a few dishes, mostly appetizers that always stay on the menu. The only entrée that I can think of that never leaves is the Steak Frites, because French fries are always in season."[4]

The layers of the salad ensure a full complement of flavors in each bite. *Photo by Clarissa Johnson.*

Two frogs, some fresh vegetables and a potato hash brown cake— sounds like the start of a bad joke. *Photo by Clarissa Johnson.*

Even when hosting special events, Terra stays true to its locavore mission. In August 2011, Terra hosted a weeklong event to honor the legendary Julia Child on what would have been her 100th birthday. Offering a three-course meal, complete with preselected wine pairings, Terra chose to honor Child's legacy and style by taking some of her classic dishes and adding a South Carolina twist.

The first course featured a play on the classic sole à la dieppoise, using fresh South Carolina flounder in place of the sole. The second course was caneton aux peches—or, for those of you non-French speakers, duck with peaches. Of course, the local tribute here came from the use of South Carolina peaches. Remember that this dinner was held in mid-August, so those peaches were about as fresh as could be. Contrary to popular belief, South Carolina actually has a higher level of peach production than Georgia, the self-proclaimed Peach State, though which state produces the tastiest peaches is still up for debate.

Given the quality of food and the dedication to a locavore style, it should not come as a surprise that Chef Davis and Terra have had an impact both in Columbia and on a national scale. As an advocate of using local food, Davis has appeared on SCETV public television to explain the importance and impact of a local, sustainable lifestyle. Davis had quite the year in 2009, when he cooked for the annual Southern Foodways Alliance benefit, Taste of the South and at Blackberry Farm and was featured as *Restaurant Hospitality*'s September 2009 Rising Star. In early 2010, Davis was invited to the highly revered James Beard House in New York City to cook his Southern Roots Dinner for a sold-out crowd. Davis has also been interviewed twice for the NPR program *A Chef's Table* and was featured in the April 2011 issue of *Garden and Gun* magazine.[5]

With partnerships in the local community and the support of those seeking out a local dining experience, Chef Davis has established himself and Terra as leaders in Columbia's local food scene. Staying true to his southern roots, Davis has been able to take classic cuisine and spin it into a vibrant take on local traditions and ingredients.

Chapter 13

COME AND KNOCK ON OUR DOOR

If you own or co-own one very successful restaurant, it is an impressive feat. If you own and run two popular restaurants, it is extraordinary. If you own and operate three successful restaurants in the same city, you are a force to be reckoned with. Enter Chef Ricky Mollohan, executive chef and owner of Columbia staples Cellar on Greene and Solstice Kitchen and Wine Bar and co-owner of Mr. Friendly's New Southern Café. What makes his story even more remarkable is that Chef Mollohan is self-taught. No fancy culinary schools or internships. Mollohan is simply someone who worked his way up the ladder, climbing from part-time employee to full-time owner.

It all started during his sophomore year at the University of South Carolina. After spending one year bouncing around assorted Columbia restaurants, learning the basics, from front-of-the-house training to knowing his way around the kitchen, Mollohan found himself knocking on the door of Mr. Friendly's. While he finished up his degree, Mollohan waited tables, managed the restaurant and worked in the kitchen. By the time graduation hit, Mollohan had a huge stake in Mr. Friendly's, and shortly thereafter, when the opportunity presented itself, he became part owner of the restaurant. Since then, Mollohan has spent his time perfecting his craft, learning alongside several accomplished chefs from across the United States.

In 2005, the second of Mollohan's major contributions to the Columbia food scene opened when Solstice claimed its niche in the northeast part of town. Though similar in mission and obviously similar in management

Worcestershire/brown sugar grilled pork tenderloin with jalapeño-smoked gouda pimento cheese, Vidalia onion/bourbon barbecue sauce, red beans and rice and collards at Mr. Friendly's. *Photo by Clarissa Johnson.*

style, Solstice set itself apart from Mr. Friendly's by catering to a mostly upscale crowd and focusing on creating the best possible dishes using local ingredients. Whereas Mr. Friendly's takes a more traditionally southern approach to its dishes, Solstice embodies specialties from around the country and adds a local twist.

By 2008, with both Mr. Friendly's and Solstice holding their own in the Columbia restaurant world, Mollohan took a chance on buying the space two doors down from Mr. Friendly's and opening Cellar on Greene. Sharing a focus on promoting foods from local vendors, Cellar on Greene sets itself apart by operating a wine store by day and a wine bar by night.[1]

Step into Mr. Friendly's on any given day, and you can expect to find families out for an evening meal, couples enjoying a date or local businessmen and women conducting meetings or enjoying a leisurely break from the office. The crowd is diverse, and the restaurant is always full, a true testament to its status as a Columbia mainstay. From its early beginnings as a sandwich and cookie shop in the early 1980s, Mr. Friendly's has embraced the local spirit of Columbia.[2] By 1995, Mr. Friendly's had

The zucchini fritters at Cellar on Greene make a wonderful option for dinner or an appetizer for two. *Photo by Clarissa Johnson.*

reinvented itself to showcase the freshness and flavors of ingredients provided by South Carolina farmers and vendors. Using simple, southern-style preparations, Mr. Friendly's prides itself of the diversity of its menu and the incorporation of products from local purveyors like City Roots, Wil-Moore Farms and Prestige Farms.

While seasonal eating factors play a large part in determining what exactly goes on the menu at Mr. Friendly's, there are some things that remain constant. "All three of our menus have their classic, customer-requested, year round dishes. I make sure these dishes are consistent and always available. But 90% of our menus change based on seasonal ingredients," said Chef Mollohan.[3] At Mr. Friendly's, one of those unchanging dishes is its famous bacon-wrapped Angus filet mignon with "French Quarter" pimento cheese. Served with sides of roasted garlic mashed potatoes and sautéed vegetables,[4] the dish is a favorite for a reason. And if you have never had the pleasure of trying Mr. Friendly's pimento cheese, you should probably stop reading this book (or take it with you) and find your way to Mr. Friendly's immediately to try it.

At Solstice, you certainly will not see pimento cheese on the menu except as an accompaniment to the fried green tomatoes, but you will see some fancier spins on traditional food. With entrées ranging from filet mignon to shrimp and grits and appetizers showcasing everything from foie gras to different varieties of oysters—as well as, of course, Solstice's signature side, smoked gouda and applewood bacon mac and cheese—those of you with a slightly more refined palate will find something to satisfy your cravings at Solstice. To further entice diners, both Solstice and Mr. Friendly's list all of their local vendors on their menus. Since South Carolina cannot provide everything the two kitchens need, Mollohan makes it a point to outsource to other vendors who support sustainable methods like Honolulu Fish Company and the Wild Salmon Company.[5]

Walk into Cellar on Greene, and you will be greeted not only by a very friendly and knowledgeable staff but also by floor-to-ceiling shelves and racks of wine. Think the library in Disney's *Beauty and the Beast*, but with wine instead of books. It may seem like odd décor for a restaurant, but Cellar on Greene is primarily a wine shop. During the day, while nearly next-door neighbor Mr. Friendly's is serving lunch to Columbia's working professionals, Cellar on Greene is selling the wine that those same professionals will take home to enjoy with their significant others. Once the clock strikes 5:00 p.m.—closing time for most businesses—Cellar on Greene opens its doors to serve the dinner crowd. With a casual wine

Lobster mac and cheese at Cellar on Greene. *Photo by Clarissa Johnson.*

bar kind of atmosphere, complete with a giant communal seating table, couches and armchairs, the patrons are usually relaxed and comfortable even before the wine starts flowing. Featuring a full bar with an extensive wine list, Cellar on Greene offers food for anyone's appetite. There is a nightly three-course meal option, separate entrées, appetizers, salads, pizzas and desserts, all of which pair wonderfully with one of Cellar on Greene's featured wines (not sure which one? Just ask the staff. They are always happy to give recommendations).[6] Regardless of what you decide on, the seared rare yellowfin tuna nachos are a sure bet for good food.

Regardless of which of Chef Mollohan's three establishments you choose, you can rest assured that you are getting some of the best ingredients Columbia has to offer. Because all three focus on seasonal eating, you will be able to choose something different each time you go. And you can feel confident that each restaurant is giving back to the community by using local vendors such as City Roots, Manchester Farms, Caw Caw Creek Farms and Wil-Moore Farms.

Chapter 14

COMMUNITY DINING IN AN UNLIKELY SPOT

Tucked away in the corner of a busy Forest Acres shopping center, on first glance Rosso does not really scream "neighborhood eatery." From the outside, it may look a bit too fancy for its Forest Acres location. But then you go inside. Here, the décor matches the atmosphere—it is intense yet not intimidating. "Alluring" might be the best word for it. Rosso just has this certain atmosphere that suggests something wonderful is about to happen.

The brainchild of Columbia Renaissance man Kristian Niemi, Rosso got its start in 2009 after Niemi sold his shares of Mr. Friendly's and Solstice. Throughout his restaurant career, Niemi really tried to embrace the local food aspect, but sometimes it was more difficult than expected. In Rosso, he had the chance to really integrate the desired local products with rustic Italian cuisine. "With Rosso, I envisioned a traditional Italian restaurant that relied on local, seasonal items to populate the menu. While it's still a bit of a challenge, every month brings to light new producers we can cultivate relationships with," he said.[1]

If anyone can do it, Niemi can. He has been working steadily in the culinary field since leaving his position as a Farsi translator for the army. During a brief stint at the College of Charleston, Niemi got his culinary start, working at Ferante's. Shortly thereafter, he left his pre-med program and transferred to the University of Minnesota, where he ultimately earned a degree in the historic preservation of architecture. While working on his degree, he enrolled concurrently at the St. Paul College for Culinary Arts. At the same time, he worked under the tutelage of Chef Ken Goff at the

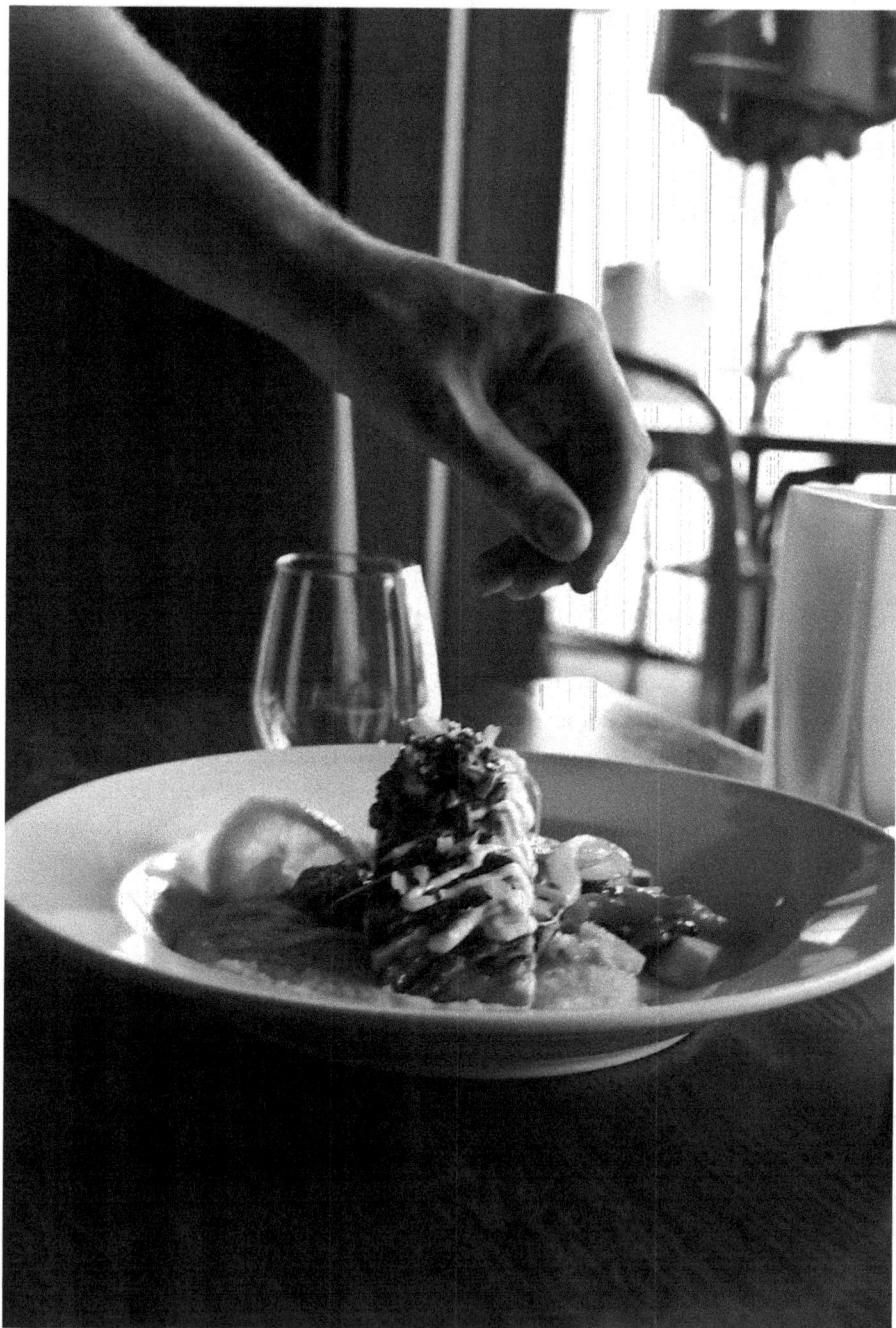

Putting the finishing touches on dinner at Rosso. *Photo by Clarissa Johnson.*

Dakota Bar & Grill, an establishment that made a name for itself as one of the first high-end, Midwest-oriented restaurants.

After finishing school, Niemi headed back south and settled in Columbia, where he immediately launched himself back into the restaurant world, first as a bartender at Garibaldi's, then as a manager at Longhorn Steakhouse and ultimately as the opening general manager of Blue Marlin. These varied experiences prepared him well for the day, a few months later, when he partnered with his floor manager and assistant kitchen manager to purchase Mr. Friendly's in early 1995. At that point, Mr. Friendly's evolved from a sandwich shop into the style of food that Columbians know and love today. "We completely changed the menu and focus of the restaurant to something more similar to what I had done at The Dakota, but with a Southern twist," said Niemi. "I instituted the most progressive wine program in town and tried valiantly to use local produce, but it was nearly impossible at that time. There were very few farmers who catered to restaurants at that time."[2] A few years later, Niemi again tried to cultivate a locavore-like atmosphere when he opened Gervais & Vine in 1999. Like with Mr. Friendly's, it was a struggle to get the local farmers to distribute to the restaurant, but Niemi did not quit.

In 2005, Niemi teamed up with local farmer and entrepreneur Emile DeFelice to host a brand-new all-local farmers' market at the restaurant once a month. The market ended up being much more popular than expected, and it ultimately led to Niemi's introduction to local farmers who were willing and able to supply his restaurants with produce on a seasonal basis. The market grew exponentially, moving from one Saturday each month to a weekly meeting. As the popularity grew, so did the need for more space, and the market commandeered 701 Whaley Street each Saturday. At the end of 2012, the market and its ever-increasing list of vendors decamped to Main Street to form the new Soda City market.

With a hand in just about everything having to do with local food and sustainability—he has even started making soaps out of the leftover fats from Rosso's kitchen—Niemi has solid reasons for focusing on the local experience. "I do all this for the same reasons anyone else in the nation does it. Fresh produce is fresher when it travels fewer miles and in less time to the plate. Diners benefit from better, tastier food, and local farmers benefit from being paid properly for it. It helps the local economy. It benefits our guests. It's important to know where your food comes from, but most people don't have the time to do it. We do it for them, and we love doing it."[3] It appears that love is returned. Scouring websites like Yelp, OpenTable and

Urbanspoon leads to utter raves about Rosso. One OpenTable reviewer called it "some of the freshest and most creative food in Columbia."[4]

The emphasis on local, seasonal and sustainable food is evident in each menu item at Rosso. The menu may not be as extensive as it is at some other establishments, but the care taken with each dish is evident in the method of preparation, plating and, ultimately, in taste. Partnerships with City Roots, Wil-Moore Farms, Caw Caw Creek and others really drive home the message that Rosso is a place dedicated to serving and being a fixed part of the local community.[5] The question, then, becomes whether or not local diners can always count on having a place like Rosso to meet and eat. If Niemi has his way, he at least will be dedicated to the local atmosphere. "I'll continue to seek out and promote local food as long as I own restaurants in Columbia, and use some of our profits to benefit local charities," he said. "It's what gets me out of bed in the morning."[6]

Chapter 15

HE CAME UP THE GULF STREAM
FROM SOUTHERN WATERS

What happens when you take the influence of European ancestry, Cajun and Creole spices and the scents and flavors associated with the Gullah and mix them all together? You wind up with pure Lowcountry flavor. Though Columbia is a few hours from the start of the Lowcountry, diners can experience the distinct taste of the region with a trip to the Vista and Blue Marlin.

The Lowcountry, for all of you non–South Carolinians, is the region of the state along the southern coast. Depending on whom you ask, the exact boundaries of the Lowcountry vary. The area in and around Charleston is usually accepted as both the largest economic center and the northernmost point. The region then spreads down the coast, including those almost coastal but still a bit inland counties, until reaching Savannah, Georgia. Culturally, the region is largely affected by its former plantation history and the influence of both European owners and African slaves.[1]

Why does this matter? Well, because Blue Marlin, though geographically a Midlands-area establishment, specializes in the traditional food of the Lowcountry, with a bit of a local twist. The flavors might come from southern South Carolina (with a hint of the Louisiana delta), but many of the ingredients are procured from vendors in the Midlands.

As you would expect from a restaurant called Blue Marlin, the primary menu offerings center on seafood—again with the Lowcountry influence. The coastal region is a hotbed of fresh seafood, with everything from oysters to catfish living in the marshes and the ocean. As with most seafood, using too much of it, or using it in the wrong way, can create major problems in the food chain. Blue Marlin has found a way to alleviate this issue by partnering with the South Carolina Aquarium's Sustainable Seafood Initiative.

The Sustainable Seafood Initiative is on a mission to make sure that the fish used in restaurants has been farmed in a way that promotes conservation and preserves the resources that the ocean provides. On its website, the Sustainable Seafood Initiative defines sustainable seafood as "fish that are caught or farmed with consideration for the long-term viability of individual marine species and for the oceans' ecological balance as a whole."[2] In late 2002, the initiative began by establishing partnerships in the Charleston area. It encouraged local chefs to verify the sustainability practices of their vendors and to increase the use and availability of local seafood on their menus. Blue Marlin has certainly embraced this philosophy, noting on its website, "We believe that enjoying fresh seafood should come with the knowledge and understanding that we must take care to consider the long term viability of the species and to provide in-season options to prevent overfishing. We make every effort to provide our guests with the best quality seafood and in doing so we buy our seafood sustainably, domestically, and locally."[3]

Partnering with the Sustainable Seafood Initiative provides Blue Marlin with an easy way to make sure it is, in fact, providing local sustainable seafood to its customers. As part of the partnership, the Sustainable Seafood Initiative reviews menus to determine sustainability. The assessment evaluates, among other criteria, stock status, ecosystem impacts, government regulations and production system to make sure that the seafood in question matches acceptable sustainability standards, while at the same time giving a basic lesson in sustainability to chefs and restaurants.[4]

While the sustainability aspect is obviously huge for a seafood restaurant, the question then becomes whether or not Blue Marlin applies the same sort of reasoning to its other ingredients. The answer is a resounding yes. In another partnership with yet another local organization, Blue Marlin works directly with Certified South Carolina Grown through its Fresh on the Menu program. As part of the Fresh on the Menu program, participating chefs and restaurants design their menus to incorporate at least 25 percent Certified South Carolina Grown foods during each season.

Certified South Carolina Grown is, by its own definition, "a cooperative effort among producers, processors, wholesalers, retailers and the South Carolina Department of Agriculture to brand and promote South Carolina grown produce and products."[5] Fresh on the Menu is the second part of the Certified South Carolina program and got its start in early 2008. As with all the various pieces of the program, the primary goal is to use and develop agricultural commerce channels to positively affect the economy of rural South Carolina. In order to achieve that goal, Certified South Carolina has created a network to connect consumers with all of the various outlets supporting Certified South Carolina, from farmers to markets to chefs.

Blue Marlin is no stranger to using local ingredients. In fact, its famous grits (ask anyone in Columbia what to order at Blue Marlin, and the answer will be shrimp and grits) have always come from literally right across the road at the Adluh Mill. The restaurant has proudly taken a stand to support other local businesses, and as a result, it is able to provide diners with ultrafresh produce and ingredients.

Of course, the ingredients, from seafood to produce, lend themselves perfectly to the Lowcountry cooking style for which Blue Marlin is famous. Drawing on an old legend for inspiration, the chefs at Blue Marlin have been able to combine the best of what both the Midlands and the Lowcountry have to offer. "Legend holds that the wealthy English, German and French plantation owners along the Carolina and Georgia coasts preferred a blander European cuisine, but while they dined in all their elegance, the aroma of Lowcountry flavors coming out from the back kitchens could not be ignored. This simple fare, created from what the good earth provided, is still a hallmark of the Southern kitchen some three hundred years later. We are proud to bring the flavors of this fare to you at the Blue Marlin."[6] That simplicity in style, though certainly not in flavor, is what has allowed the local Midlands ingredients to gel so well with the flavors and traditions of the Lowcountry.

Chapter 16

KEEP YOUR NOSE TO THE GRINDSTONE

G rits are one of those foods that just scream *southern* whenever they are mentioned. Served as a part of breakfast, as an accompaniment to lunch and as a base for dinner, grits have an amazing range of versatility. And while there are not quite as many types as there are uses for grits, there is still a variety to suit every situation and every palate. For those in a hurry, there are infinite types of instant grits available at any local grocery store. There are also quick-cooking, non-instant grits and long-cooking grits. And in South Carolina, there are Adluh grits. No, Adluh grits are not some magic brand of grit designed to give superpowers to the eater. They are grits that are milled and ground in the very heart of Columbia, made from South Carolina corn. Does it get much more local than that?

Adluh has been running as a flour mill in the Vista since about 1900. For about twenty years, the mill was owned and operated by B.R. Crooner and his family. Then, in 1920, Adluh joined forces with J.H. Hardin's Columbia Grain and Provision Company, and the mill ran under joint control until 1926, when the site went into foreclosure. A family from North Carolina, the Allens, purchased the mill from the bank in 1926 and have kept in in the family ever since, a fact that is not really surprising given that the Allens have owned and operated a handful of gristmills in North Carolina since the early 1800s. The Allens were successful in their business endeavors and, within ten years, had moved one of their North Carolina plants to Greenwood, South Carolina, raising the number of Allen-owned mills in the state to two. Now, more than seventy-five years later, while the Greenwood mill is no longer open, the Columbia mill is still thriving.[1]

So why, when there were forty-two fully operational mills in South Carolina in 1942, is Adluh the only one that is still running? The answer is simple: quality. The quality of Adluh's products is what has kept the mill in business. Part of that can be attributed to the fact that the Allens still use the same techniques that the mill used when it first opened in 1900. The mill's website proclaims, "Without this emphasis on 'Same Today, Same Always' quality, the company could never have survived the shift in consumption from local to regional to national brands over the last 100 years. Generations of customers know that they can always trust their prized recipes to come out right using Adluh products."[2] That last bit is the key. Adluh has greatly benefited from being a Columbia icon. Recipes handed down from generation to generation, at least for the past one hundred years, have sworn by Adluh's products. From grits to flour, Adluh has built a name for itself that Columbians recognize as synonymous with good quality.

Columbia residents, and indeed all South Carolina residents, can take pride is using Adluh's products. Not only does the company produce at the highest level of quality, but Adluh is also active is supporting the community. Most of the wheat and yellow corn used in Adluh's grits and flour is grown in South Carolina in an effort to support local growers.[3] When those same growers have faced harvest issues due to drought, Adluh has partnered with the Clemson Extension Service and the South Carolina Department of Agriculture to supply animal feed to the farms. In addition, for anyone interested in how the mill operates and exactly how each food is produced, Adluh provides tours to everyone, from school groups to church groups and everyone in between. It also runs a museum, highlighting the mill's more than one-hundred-year history, which is opened to the public during the annual Congaree Vista Lights festival in December.[4]

It has also helped that although its processes have not changed much since 1900, Adluh as a brand is not afraid of keeping up with new business methods and changing technology. The most recent shift was a foray into the world of social media. Of course, adventurous marketing techniques are nothing new to the folks at Adluh. In an interview with *Columbia Business Monthly*, the company's controller, Beth Ellis, related a tale of some earlier marketing techniques that Adluh used to not only promote its products but also drive home that feeling of community and the benefits of buying local products. "We had a truck that went around town in local neighborhoods. Our customers were all home consumers. We would have people drive around and knock on the door."[5] Thus, the truck driver earned the moniker of the "Adluh Knocking Man." When he appeared on the doorstep, any customer who could show him Adluh flour in the house was given a cash prize.[6]

The days of the Adluh Knocking Man are long past, but new methods have taken effect. Building on the already-established community of loyal customers, Adluh added Internet sales to its repertoire in the early part of the twenty-first century. Through this expanded sales capacity, Adluh has been able to reach a new population: those who grew up on Adluh products but no longer live in or around Columbia. As Internet sales grew, so did the popularity of social media. By establishing a presence on both Twitter and Facebook, Adluh has found yet another way to connect with its customers.

A quick glance at Adluh's Facebook page provides a great look at just how much this company has affected the community. Historic pictures of the mill, recipes using Adluh products and general information about local restaurants featuring Adluh's products on their menus are the highlights. And those are just the pieces that Adluh posts. People who like Adluh's page have shared some of their favorite recipes, expressed their gratitude for Adluh's involvement in the community and, more importantly, noted some of the traditions that have been passed down through generations that revolve around Adluh products, including holiday baking.[7]

The argument could certainly be made that Adluh flour stops being a local product once it is shipped out of the state. It would be equally easy to argue, however, that by shipping products around the country, and by using Facebook and Twitter to connect and keep up with their customers, Adluh is actually fostering the sense of community that is so important to local eating. The old adage proclaims, "Home is where the heart is." For Adluh's patrons who no longer live in Columbia, the social media techniques and online shopping options allow them to take a little piece of home with them to keep that community going. Edward Sharpe and the Magnetic Zeros sing, "Home is wherever I'm with you."[8] While the song is a love story, this line from the chorus holds true in this case. Adluh is such an icon in Columbia and has had such an impact on so many people's lives that no matter where the customers are, that relationship to their home community endures. That sense of community, to paraphrase Linus in *A Charlie Brown Christmas*, is what local eating is all about, Charlie Brown.[9]

Chapter 17

MEATY BEATY BIG AND BOUNCY

E mile DeFelice is a busy man. Not only does he oversee the running of Soda City every Saturday, but he also owns and manages the one-hundred acre Caw Caw Creek farm in St. Matthews, South Carolina. Caw Caw Creek, unlike most of the farms already mentioned, is devoted not to vegetables and herbs but to pigs. Yes, pigs. Those wonderful animals that are the source of all of those delicious sausages, tenderloins, ribs and the granddaddy of them all (apologies to the Rose Bowl), bacon. Really, the variety of products is astounding. Who knew that there were so many parts of a pig that could be used for food? And those are just the normal products for everyday use. You can also get feet, skin, tails and assorted types of fat for all of your cooking needs.[1]

The need for organic and sustainable farming methods is obvious when it comes to vegetables and grains. With pigs, it is equally important, though maybe not as immediately clear as to why. But like all things sustainable, much of it has to with the effects of factory pig farming on the environment, the treatment of the pigs and the overall threat to the community. In Europe, it is such a huge issue that a team of documentarians produced a film called *Pig Business* to highlight the horrors of factory pig farming.[2] Though *Pig Business* focuses on European pig farming practices, it is important to note that the issues are the same in the United States. Caw Caw Creek is a prime example of the principles of farming that *Pig Business* encourages. To start, Caw Caw Creek's pigs are raised in what DeFelice calls a "managed wild setting,"[3] meaning that the pigs are pretty much allowed to roam free on the farm to enjoy life as pigs should—none are forced into claustrophobic pens. The pigs have access to whatever fodder they require whenever they feel like eating, natural sunlight and companionship.[4] Apparently, pigs are a lot like humans when it comes to being happy.

Per its website, Caw Caw Creek's mission is to provide the most delicious pork products it can, in ways that are good for both the pigs and the consumers.[5] So far, it looks like Caw Caw Creek has achieved its mission. Restaurants and families across the Midlands—and really even the country, thanks to an online ordering system—are the primary purchasers of Caw Caw Creek's products. The taste and quality are even more evident when you consider that some of the best chefs in the world (Daniel Boulud, Frank Stitt and Michelle Bernstein, among others) have created dishes using Caw Caw Creek's products.[6] Further proof that local and sustainable methods just create a better product.

Much like Caw Caw Creek, Wil-Moore Farms also specializes in humanely raised animals. The farm is pasture based, meaning that all of the animals are raised on Wil-Moore Farms' certified organic pasture, which is in keeping with the animals' natural diets.[7]

Located northeast of the capital city, in Lugoff, South Carolina, Wil-Moore Farms is a family affair. Owners Robin and Keith Willoughby are the third generation to call Wil-Moore Farms their own. They place a tremendous emphasis on the health benefits of eating grass-fed livestock. On the farm's website, some of these health benefits are touted. "Meat and eggs from grass-fed animals supply our bodies with a natural source of Omega 3 Fatty Acids, CLA (conjugated linoleic acid) and Beta-Carotene. Omega 3 Fatty Acids have been shown to improve the heart and may reduce your risk of cancer."[8] On its home page, Wil-Moore Farms affirms its mission to produce nutritious and tasty food. The Willoughbys also control every facet of the business, from farming to marketing, so they truly do everything they can to keep Wil-Moore Farms as local and sustainable as possible.

Wil-Moore Farms has made this commitment, and it sticks to it in all of its products. It offers assorted cuts of beef and beef sausage, pork by the cut and three different kinds of pork sausage, as well as lamb and goat cuts. And then there is the poultry. Wil-Moore Farms raises free-range poultry, which pretty much means that the chickens and turkeys are also raised on pasture. They have free run of the pasture and are provided with shelters in case they feel like they need to get out of the sun. As a result, the chickens and turkeys that are harvested are done so all naturally and humanely. With chickens, of course, come eggs. Wil-Moore specializes in brown eggs, which are all natural (meaning hormone and antibiotic free) and processed daily to provide the freshest eggs possible to consumers.[9]

There is always the taste factor to consider too. By using free-range and pasture methods, the cuts of meat end up being juicier, more tender and more robust in flavor than from animals raised using other methods. With Wil-Moore Farms, you can be sure that the animals are being treated well and harvested carefully and are free from additives or preservatives that may contribute to health issues. The farm's products are utterly delicious and nutritious.

Chapter 18

GET YOUR MOTOR RUNNIN'

As the first restaurant established in the Vista, Motor Supply Company has been setting the bar for Columbia dining since it opened its doors in 1989. There have been some changes since its inception, but one thing has always remained the same: the focus on serving a diverse blend of food. With menus that change twice daily and an emphasis on using fresh and local ingredients, Motor Supply Company has established itself at the forefront of the Columbia dining scene.

That is in no small part due to executive chef Tim Peters. Peters took control of Motor Supply's kitchens in 2005 and brought with him not only extensive wine knowledge but also a deeply ingrained love of using local, fresh and often organic products. Part of that love came out of Peters's brief time spent working on an organic farm on Wadmalaw Island of the South Carolina coast. The time on the farm allowed Peters to really understand what it means to connect with the earth, the soil and the products that result. He also discovered just how much of a struggle that kind of life can be for small, local farmers who are trying to get their products out to the masses. Peters has learned from his farmhand experiences and applies many of those lessons to the kitchens at Motor Supply. As he states in his bio on the restaurant's website, "We feel that if we stick with these products, then it will pay off for everyone in the long run."[1]

Peters's dedication to the use of local products has earned him some recognition in the community. In 2012, he was one of the ten chefs to be named Slow Food at Indie Grits 2012 Sustainable Chefs. Part of that may be due to Peters's help in starting Harvest Week at Motor Supply. For one week each season, Motor Supply features events with some of the growers and farmers who supply the restaurant with their ingredients and, of course, makes it a point to showcase some of its dishes made using those same ingredients.[2]

At Motor Supply, it is not just the ingredients and dishes that support sustainable living. Since Peters came on board, Motor Supply has really stepped up its game when it comes to getting involved in environmental efforts around the city. In an article by Tenessa Jennings, Motor Supply's owner Eddie Wales was quoted as saying, "Owning a green business in Columbia, SC is great—you can feel the enthusiasm for sustainable projects radiating from City Hall and even from the SC State House."[3] He continued, saying, "Columbia City Council passed an initiative in 2011 to pursue a 30-year zero waste policy, and the lawmakers who come down the hill for lunch & dinner at Motor are backing more green legislation than ever."[4] It does not hurt that Motor Supply is essentially in the backyard of the University of South Carolina, whose sustainability efforts in the past few years have been the stuff of legend. Motor Supply has taken full advantage of the proximity and resources of the university and contributed to the vegetable scrap composting initiative started by Sustainable Carolina and the organic herb garden started at the university's Green Quad dorm. As Wales stated in his interview with Jennings, "Being green has been such an easy choice for us, ideologically…It's just right in line with what Tim is doing with the food, buying from local, sustainable farmers."[5]

For Peters, the concept is simple. Using fresh, local food with constant variety is what allows him to provide the dining experience he does. In a 2012 interview with Discover South Carolina, Peters referred to Motor Supply as a "farm to fork bistro."[6] Part of what makes Motor Supply so exciting is its constantly changing menu. Plans for a dinner menu may change if a vendor calls to give Peters a heads up about produce that just became available. "I love the surprise visits and phone calls telling me that livestock is ready or I am about to get my hands on heirloom vegetables still warm from the sun,"[7] Peters said in his interview.

One of those suppliers making surprising phone calls is Freshly Grown Farms in Columbia. Motor Supply serves some incredible salads, and if you have ever tried one, chances are you have eaten one of the many varieties of lettuce grown by Freshly Grown Farms. It may seem a little boring to focus on lettuce, but Freshly Grown Farms uses a full-on hydroponic technique to eliminate water waste and create a positive environmental impact. This type of growing technique also eliminates the need for pesticides and other harmful chemicals. Instead, Freshly Grown Farms is able to use all-natural methods of pest control, which ultimately provides consumers with a healthier, more sustainable product. Lettuce from Freshly Grown Farms will also last for a few weeks as a direct result of its farming techniques. When it is sold and distributed to both restaurant and retail partners, the lettuce arrives with the roots still intact, which helps extend the freshness of the lettuce.[8]

The only drawback to an operation like Freshly Grown Farms is that sometimes the supply exceeds the demand, which is a great problem for the farm to have but not so great for its partners. Peters actually had to make an effort to scale back on his use of Freshly Grown Farms' lettuce in early 2012 because his constant demand for the product was hindering the farm's ability to supply other restaurants.[9] Having a chef that excited about food, especially the ingredients provided locally, really carries over into the food that is brought to the table and makes for an incredible dining experience.

Chapter 19

WHERE DO WE GO FROM HERE?

Columbia is in the midst of a food revolution that rivals that of Jamie Oliver. But the movement is still comparatively small. As Slow Food Columbia hosts and supports more and more events around town, the world will get out. The fact that City Roots has exploded in popularity in its brief existence is encouraging. More and more restaurant chefs are hopping on the local food train, eager to explore what the Midlands region has to offer. The movement will continue to grow, but only if the city's residents do their part to get the word out.

Author Tanya Denckla Cobb presented a compelling case for the local food movement, likening it to other grass-roots movements that have led to major change in the United States. In her book *Reclaiming Our Food*, Cobb explained that little by little and person by person, the citizens of this country are on a mission to combat the processed food giants and regain a connection to the land and the food. As Cobb noted, each person's reasons and experiences are different, but in the end, becoming part of this movement strengthens both a relationship with food and with community. "No matter the starting point, no matter how richly diverse the motivations or approaches…successful grassroots food projects ultimately converge around two central points: local food and community. A community can grow a more sustainable and resilient economy by growing its local food system, and a healthy local food system will nurture and grow community spirit."[1] Based on the stories included here, it would seem that Columbia is taking Cobb's advice and striving to build a community where food and culture go hand in hand.

The leaders in this movement are the people and organizations that have been the focus of this book. Each has taken the lead in some way to shape the food identity of Columbia. While they might have pioneered the cause, they are certainly not the only ones responsible for spreading the world. In a time when ecological and economic crises, as well as the oft-mentioned obesity epidemic, vie for their respective moment in the spotlight, it is important to know that everyone needs to take a little bit of responsibility toward implementing change. As Cobb brilliantly pointed out, people are now living in a time when they are so far removed from their food that most could not tell you the difference between a carrot and a parsnip.[2] Only a few generations ago, seemingly everyone at least had an idea of where their food came from. "The 'greatest generation' might remember having even fewer degrees of separation from their food production, when meat was obtained from a local butcher (and the butcher knew the farmer who had raised the animal) and vegetables came from backyard gardens or nearby farmers."[3] Today, most people could not even begin to theorize on where exactly their food had come from.

Of course, the "most people" generalization does not include those who support local and sustainable eating. In Columbia, chefs are extraordinarily diligent about noting exactly where each of their local ingredients originates. The menu will clearly mention that something is topped with City Roots microgreens or that it features pork from Caw Caw Creek. It is a small step toward breaking down the knowledge barrier but one that most Columbia locavores appreciate.

The key to any successful grass-roots movement lies in the strength of the community. To this point, Columbia has built a strong network of local food aficionados and enthusiasts, but is that enough to sustain an entire movement? Maybe. Sometimes, all that is needed is one person passionate about the movement to light a fire, and then the rest falls into place. Cobb, who calls those who take the initiative to bring local food resources out of the dark "the unsung heroes of community food movements,"[4] thinks that with the right tools and enough courage, any movement can succeed. "Where there is a will, there often is a way," Cobb noted. "A sustainable food community system requires weaving a web of connecting threads…Regardless of their size, these innovative enterprises are stimulating their local economies, growing new 'green' jobs, returning profits to local producers, circulating food dollars within local economies, and creating markets that support greater numbers of small-scale producers and farmers."[5]

In the Columbia food scene, the unsung heroes have worked tirelessly to finally get the recognition they deserve, both on the local level and, in some cases, the national level. Each has worked hard to spread the word about the importance of local and sustainable food and how Columbians can get involved. For a city that has always banded together to support its own, it seems even more crucial now that the driving force of this movement shifts away from the farmers and the chefs and becomes the purview of those who most desire to consume local products. Word of mouth is a wonderful marketing tool, one that will serve the Columbia community well as more and more people indulge their curiosity about local and sustainable foods and wonder what they can do to be a part of the movement.

GRAB YOUR THINGS, I'VE COME TO TAKE YOU HOME

Though I think I always had the knowledge buried in the back of my brain, during the writing of this book, I became reacquainted with the idea that food is shaped by the community, location and cultural background of its preparers and consumers. In the major city bordering my hometown, for example, a Sunday afternoon stroll down an array of neighborhood blocks brings with it the smell of bubbling gravy (that is spaghetti sauce to all of you non-Philadelphians) and pasta. Along with that tease to your nose, you will also see multiple generations of one family gathering to celebrate major life events or even just one another's company. My point is that food should not just be looked on as a means of staying alive. Behind every dish is a story of a group of people, a geographical location or a memory. Re-creating those feelings is my personal takeaway from the local food movement. There is a deeper connection to the earth, to the farmer and to the community than I would find in a bag of mass-produced frozen peas.

Ask anyone who has ever spent any time here, and they will tell you that South Carolinians are ridiculously proud of their heritage. For Columbians, that heritage is often an amalgamation of cultural influences. If they were lucky enough to grow up in Columbia, there is probably a bit of pride in the way Columbia always seems to pick itself back up, even when its counterparts in the state are struggling to do so. There may be a knowing smile when an outsider asks what in the world a Crowder pea is (trust me, I have definitely experienced that one). They may explain to you as you drive along a street or a section of highway named for one of their ancestors why he was so important to South Carolina history.

South Carolina has always been a state whose legacy is intertwined with the fate of local farming. From the Upstate, through the Midlands and into the Lowcountry, local food has shaped diets as well as communities. On a national scale, locavorism and sustainability have taken over the conversations of foodies, farmers, lobbyists and environmentalists. With concerns about greenhouse gas emissions, "food miles," land quality, the economy and a finite amount of natural resources, people across the country have decided to decrease their collective carbon footprints and support their communities at the same time. The increased availability of locally and sustainably grown products—even for those needing to use food assistance programs—has really catapulted locavorism into the spotlight.

Even Columbia has returned to its farming roots and jumped on the locavore bandwagon. Chefs like Mike Davis, Ricky Mollohan, Kristian Niemi and Tim Peters are leading the way by making a point to use local vendors and feature local ingredients in their dishes. City Roots has single-handedly brought about a resurgence in urban farming. Its produce can be found in restaurants throughout the state and on dinner tables across the Midlands. The farm's sustainable practices landed it a grant from the United States Department of Agriculture to further explore different farming methods. Even Pinckney's Produce, a family-owned company in Beaufort, has gotten in on the Columbia food scene. Its CSA boxes have been so popular that it has two drop-off days scheduled in Columbia during each season. The variety of fresh vegetables and more unique items like herbs has earned Pinckney's raves from Columbia's residents. And of course, what food movement would be complete without mention of farmers' markets? From the former All Local Market on Whaley Street to Soda City and the fledgling Vista Marketplace, the wares of local artisans, craftsmen and farmers have found a home.

Above all else, however, each of these organizations, restaurants and markets has led the charge to foster a deeper sense of community in Columbia. Aiding them in their quest are the patrons of their establishments. For what good is a well-produced or locally grown product without someone to purchase it? (Let's not forget that cultivating the local economy is a key part of both locavorism and sustainability.) Among those who frequent these places is a community within the community, local food bloggers. As one of them, I feel that it is important to emphasize why each of us has pursued this path. There are clear benefits to local eating, but as with all kinds of food, it is often more personal than that. Columbia has a tight-knit community of local food bloggers, some of whom share their recipes and some who, like myself, explore the dining and local cultural aspects of the food world.

Since this book was conceived with the idea of promoting the local culture, I would be remiss if I only included opinions and initiatives by those who have become the face of the local food movement—the chefs and business owners. As a result, and in an effort to include as many people as I could, I polled my Facebook followers and fellow food bloggers about why local eating and sustainability were so important to them. Because really, how can you cultivate a local food community if the local residents do not buy into the hype?

Polly Thompson, author of "La Cucina di Paulina," was quick to offer her reasons for eating locally. "Eating locally is not only generally healthier, but it supports the local economy, usually owner/operator businesses," she said. "Using locally grown and sustainable food is preferable to me because I know where and how things were grown, that the food is fresh, and promotes the environment."[1] For local resident Walter Brooker, the taste is one of the primary reasons that he supports buying locally. "Home grown always tastes better. I would rather buy local than commercial if I can help it," he said.[2]

For Lydia Scott, the decision is more about the economic implications. "Big box business is part of what's wrong in this world. The local business owners are people just like me and some of them are friends...I want them to succeed!"[3] Remsy Munib, a transplant from California, where the locavore lifestyle first gained publicity, agreed. As he said, local food "works the local economy."[4] Munib has also found that "foodie cities" like Charleston, Charlotte and Atlanta are just too far away to be frequented. As a result, he has embraced the local eateries in Columbia.

Bach Pham, who until recently authored "The Foraging Foodie," likes the personal aspect of local eating. "It's a very personal relationship, eating food that's locally grown and produced. Taking a moment to step back and get to know the people who took the time and effort to make or even grow the food that I eat day in and day out creates this special connection that really enhances my whole eating experience," he said.[5] And while the personal experience is certainly an important connection to make, Pham also acknowledged the economic benefits of eating locally. "The fact that your money goes straight to the person who grew, harvested, and cared for your produce only makes that more special. And it is not just farmers when we're talking local, but all the people who work to create amazing locally made cuisine in our restaurants and our markets, all contributing towards the betterment of our community."[6] The crowning factor in Pham's decision to eat locally? Taste. "There are few things better than a freshly plucked heirloom tomato in the summer, or a super sweet South Carolina peach pulled just down the road, mere hours before."[7] I'd have to say that I agree. Of course, I am one of the few people I know who will willingly eat fresh tomatoes like apples, so I may be a bit biased in my opinion of off-the-vine produce.

For me, the decision to eat locally is an easy one. Sure, sometimes it means spending a little extra money to get high-quality produce and award-winning restaurant meals, but I feel that the extra expense is worth it because I know that it is going right back into supporting the local vendors and restaurants I love. Do I always eat locally? Nope. I am definitely of the wild card locavore mentality—I cannot live without coffee (and trust me, you do not want to see me when I am undercaffeinated) or other goodies that are not local to Columbia. But when it comes to vegetables and meats, I do try to make it a point to give my business to the myriad local growers we have in town as often as I can.

But like with so many others, for me it is not just about the food. It is about the sense of community and belonging that local eating creates. For me, the story surrounding the food is just as important as the food itself. I have always thought of eating as a social event. Think about major milestones in your life. How many of them have featured celebrations centered on food? Super Bowl parties, football tailgates, birthday parties, baby showers and even wedding receptions. Each event brings people together over some sort of deliciousness. Food can be an easy conversation starter ("Oh my goodness, this dip is so good! How did you make it?") or a way to establish a connection to someone else ("I caught these fish this morning." "I love fish! Have you ever tried fish fingers and custard?"). Going out to eat and sitting at the bar or at a communal table will almost force you to have a conversation with someone you would not have met otherwise. If you are lucky, you may even get one of those locally famous chefs to regale you with stories of their food experiences. I find that you can learn so much about people and traditions just by stopping at the farmers' market on Saturday morning. I've been exposed to foods I had never eaten before (collards, anyone? Remember, I grew up north of the Mason-Dixon line) and have been reintroduced to foods I thought I hated (beets) just by exploring the local community and patronizing those establishments that serve the best and freshest food Columbia has to offer.

I am sure you have heard the expression "It takes a village to raise a child." In this case, it takes a city to create the food community. Bach Pham hit the nail on the head when asked about what constitutes local food: "It's not just farmers when we're talking local, but all the people who work to create amazing locally made cuisine in our restaurants and our markets, all contributing towards the betterment of our community. It's just a really wonderful network that continues to grow deeper and deeper as more people step up and help make our city a little bigger and brighter each day."[8] For a city that is often bypassed or overlooked in favor of ones with brighter lights or flashier reputations, it should be a comfort to its residents that Columbia is fostering this type of community spirit. And as a fan of all things food related, it comforts me to know that like Thanksgiving dinner at my mom's house, spending a morning at the farmers' market or an evening at one of Columbia's many local and sustainable restaurants feels like coming home.

NOTES

Chapter 1

1. Moore, *Columbia and Richland County*, 1.
2. Ibid., 2.
3. Harrelson, *Handbook of South Carolina*, 103.
4. Moore, *Columbia and Richland County*, 5.
5. Ibid., 7.
6. Ibid., 8.
7. City-Data, "Columbia: History."
8. Moore, *Columbia and Richland County*, 10.
9. Ibid., 16.
10. Infoplease, "South Carolina."
11. City-Data, "Columbia: History."
12. Ibid.
13. Moore, *Columbia and Richland County*, 77.
14. Ibid., 182
15. Ibid, 231
16. South Carolina State Library, "Brief History of South Carolina."
17. Moore, *Columbia and Richland County*, 295.
18. Ibid., 298.
19. *State*, May 23, 1919, quoted in Moore, *Columbia and Richland County*, 320.
20. City-Data, "Columbia: History."
21. Moore, *Columbia and Richland County*, 338.

22. Ibid., 340
23. City-Data, "Columbia: History."
24. Moore, *Columbia and Richland County*, 390.
25. Bureau of Labor Statistics, "Unemployment Rates."
26. Google Public Data, "Unemployment Rate—Not Seasonally Adjusted."
27. Moore, *Columbia and Richland County*, 390.

Chapter 2

1. South Carolina Department of Agriculture, "SC Agribusiness."
2. Opala, "South Carolina Rice Plantations."
3. Ibid.
4. Discover Charleston, "South Carolina Gold Rush."
5. Carolina Gold Rice Foundation website.
6. Discover Charleston, "South Carolina Gold Rush."
7. Barna, "Anson Mills."
8. Colicchio, "Day 1."
9. Anson Mills website.
10. Hess, *Carolina Rice Kitchen*, 3.
11. Anson Mills website.
12. Ibid.

Chapter 3

1. City of Columbia, "Brief History of Columbia."
2. Carolana, "History of Columbia, South Carolina."
3. Ibid.
4. Columbia Convention and Visitors' Bureau, "Signature Events."
5. Ibid.
6. Jennings, "26th Annual Chili Cook Off."
7. Rosewood Crawfish Festival website.
8. Irmo, South Carolina, "Irmo Okra Strut."
9. Ibid.
10. Lexington County Peach Festival website.

11. Columbia Convention and Visitors' Bureau, "3rd Annual Palmetto Tasty Tomato Festival."
12. Ibid., "5th Annual Viva La Vista."
13. South Carolina Oyster Festival Facebook page.
14. Harvest Dinners Series website.
15. Slow Food Columbia website.
16. Ibid.

Chapter 4

1. DeWeerdt, "Is Local Food Better?"
2. Trimarchi, "What Are Locavores?"
3. Ibid.
4. California Department of Social Services, "CalFresh Program."
5. South Carolina Department of Social Services, "Supplemental Nutrition Assistance Program."
6. Farm to School website.
7. Ibid.
8. South Carolina Farm to School website.
9. Trimarchi, "What Are Locavores?"
10. Ibid.
11. Maiser, "10 Reasons to Eat Local Food."

Chapter 5

1. Miriam-Webster, "Sustainable," www.merriam-webster.com/dictionary/sustainable.
2. Agricultural Sustainability Institute, "What Is Sustainable Agriculture?"
3. Sustainable Midlands website.
4. University of South Carolina, "Sustainable Carolina."
5. Ibid.

Chapter 6

1. Kolar, "Historical Reflections."
2. Grace Communications Foundation, "Local & Regional Food Systems."
3. Ibid.
4. Ibid.
5. Ibid.
6. Ibid.
7. Ibid.
8. Ibid.
9. Joni Mitchell, "Big Yellow Taxi," Warner Brothers, 1970.
10. Grace Communications Foundation, "Local & Regional Food Systems."
11. Jones, "Live Aid 1985."
12. Dunkle, "Willie Nelson on the Farm Aid Cause."
13. Farm Aid, "Farm Aid: Family Farmers."
14. Ibid.
15. Ibid.
16. Dunkle, "Willie Nelson on the Farm Aid Cause."
17. Farm Aid, "Farm Aid: Family Farmers."
18. Ibid., "Farm Aid's 2012 Grants."
19. Ibid., "Helping Family Farmers Thrive."
20. Steffey, "Oh Farm Bill."
21. Farm Aid, "Helping Family Farmers Thrive."
22. United States Department of Agriculture, "Regional Food Hubs."
23. Grace Communications Foundation, "Local & Regional Food Systems."
24. Ibid.
25. Ibid.
26. Ibid.
27. Ibid.
28. Sam Cooke, "A Change Is Gonna Come," RCA Victor, 1963.

Chapter 7

1. City Roots website.
2. McClam, "City Roots."

3. Robbie McClam, e-mail message to the author, January 3, 2013.
4. Ibid.
5. Ibid.
6. Ibid.
7. Ibid.
8. Ibid.
9. Ibid.

Chapter 8

1. Local Harvest, "Community Supported Agriculture."
2. Katie Thompson, e-mail message to the author, January 4, 2013.
3. Ibid.
4. April Blake, Facebook message to the author, January 6, 2013.
5. Ibid.
6. Ibid.
7. Katie Thompson, e-mail message to the author, January 4, 2013.

Chapter 9

1. Emile DeFelice, e-mail message to the author, December 27, 2012.
2. Ibid.
3. Kearns, "Farmers Market."
4. Soda City Market website.
5. Emile DeFelice, e-mail message to the author, December 27, 2012.
6. Pioreschi, "Farmers Market Huge Win."
7. Bach Pham, Facebook message to the author, January 6, 2013.
8. Frank Adams, Facebook message to the author, January 6, 2013.

Chapter 10

1. Vista Marketplace website.
2. Frank Adams, Facebook message to the author, January 7, 2013.
3. *Empire Records*, Warner Brothers.
4. Ibid.
5. Local Harvest, "Organic Food."
6. Ibid., "D&J Farm."
7. Round River Farms website.
8. Ibid.
9. Doko Farm website.
10. University Creative Services, "USC Alum."
11. Local Harvest, "Doko Farm."

Chapter 11

1. Rosewood Market website.
2. Basil Garzia, e-mail communication to the author, January 22, 2013.
3. Rosewood Market website.
4. Eison, "Basil Garzia and Rosewood Market & Deli."
5. Ibid.
6. Ibid.
7. Basil Garzia, e-mail communication to the author, January 22, 2013.
8. Eison, "Basil Garzia and Rosewood Market & Deli."

Chapter 12

1. Mike Davis, e-mail message to the author, January 4, 2013.
2. Ibid.
3. Ibid.
4. Ibid.
5. Terra website.

Chapter 13

1. Ricky Mollohan, e-mail message to the author, January 7, 2013.
2. Rely Local, "Mr. Friendly's."
3. Ricky Mollohan, e-mail message to the author, January 7, 2013.
4. Mr. Friendly's New Southern Café website.
5. Solstice Kitchen & Wine Bar website.
6. Cellar on Greene website.

Chapter 14

1. Kristian Niemi, e-mail message to the author, December 27, 2012.
2. Ibid.
3. Ibid.
4. OpenTable, Rosso Trattoria Italia.
5. Rosso Trattoria Italia website.
6. Kristian Niemi, e-mail message to the author, December 27, 2012.

Chapter 15

1. Bopp, "Road Trip."
2. Sustainable Seafood Initiative, "Sustainable Seafood Initiative Fact Sheet."
3. Blue Marlin Steaks & Seafood, "Sustainable Seafood."
4. Sustainable Seafood Initiative, "Sustainability Criteria."
5. Certified South Carolina, "Fresh on the Menu."
6. Blue Marlin Steaks & Seafood, "The Blue Marlin Story."

Chapter 16

1. Adluh Flour, "History."
2. Ibid.
3. Ibid.

4. Ibid., "Agriculture."
5. Worthy, "Adluh Flour Explores Social Media."
6. Ibid.
7. Adluh Flour Facebook page.
8. Edward Sharpe and the Magnetic Zeros, with Alex Ebert (writer), "Home," Vagrant Records, 2009.
9. *A Charlie Brown Christmas*, Warner Brothers.

Chapter 17

1. Caw Caw Creek website.
2. Pig Business website.
3. Caw Caw Creek website.
4. Ibid.
5. Ibid.
6. Ibid.
7. Wil-Moore Farms website.
8. Ibid.
9. Ibid.

Chapter 18

1. Motor Supply Company Bistro, "Bios."
2. Jennings, "Motor Supply Co. Bistro."
3. Ibid.
4. Ibid.
5. Ibid.
6. Fowler, "In the Kitchen."
7. Ibid.
8. Freshly Grown Farms website.
9. Fowler, "In the Kitchen."

Chapter 19

1. Cobb, *Reclaiming Our Food*, 8.
2. Ibid., 40.
3. Ibid.
4. Ibid., 242.
5. Ibid.

Conclusion

1. Polly Thompson, Facebook message to the author, January 28, 2013.
2. Walter Booker, Facebook message to the author, January 28, 2013.
3. Lydia Scott, Facebook message to the author, January 28, 2013.
4. Remsy Munib, Facebook message to the author, January 28, 2013.
5. Bach Pham, Facebook message to the author, January 28, 2013.
6. Ibid.
7. Ibid.
8. Ibid.

BIBLIOGRAPHY

Adluh Flour. "Agriculture." www.adluh.com/agriculture.htm.

———. "History." www.adluh.com/history.htm.

Adluh Flour Facebook page. www.facebook.com/Adluh?fref=ts.

Agricultural Sustainability Institute. "What Is Sustainable Agriculture?" www.sarep.ucdavis.edu/sarep/about/def.

Anson Mills. http://ansonmills.com.

Barna, Stephanie. "Anson Mills." *Charleston City Paper.* http://www.charlestoncitypaper.com/charleston/anson-mills/Location?oid=3629025.

Blue Marlin Steaks & Seafood. "The Blue Marlin Story." www.bluemarlincolumbia.com/the-blue-marlin-story.

———. "Sustainable Seafood." www.bluemarlincolumbia.com/sustainable-seafood.

Bopp, Suzanne. "Road Trip: Low Country, South Carolina and Georgia." *National Geographic Traveler.* http://travel.nationalgeographic.com/travel/road-trips/low-country-south-carolina-georgia-road-trip.

Bureau of Labor Statistics, United States Department of Labor. "Unemployment Rates for Metropolitan Areas." January 30, 2013. www.bls.gov/web/metro/laummtrk.htm.

California Department of Social Service. "CalFresh Program." www.dss.cahwnet.gov/foodstamps.

Carolana. "A History of Columbia, South Carolina." www.carolana.com/SC/Towns/Columbia_SC.html.

Carolina Gold Rice Foundation. www.carolinagoldricefoundation.org.

Caw Caw Creek. http://cawcawcreek.com.

Cellar on Greene. www.cellarongreene.com.

Certified South Carolina. "Fresh on the Menu." www.certifiedscgrown.com/freshonthemenu.

A Charlie Brown Christmas. Directed by Bill Melendez, 1965. Burbank, CA: Warner Brothers, 2000. DVD.

City-Data. "Columbia: History." www.city-data.com/us-cities/The-South/Columbia-History.html.

City of Columbia. "A Brief History of Columbia." www.columbiasc.net/columbia/257.

City Roots. http://cityroots.org.

Cobb, Tanya Denckla. *Reclaiming Our Food: How the Grassroots Food Movement Is Changing the Way We Eat.* North Adams, MA: Storey Publishing, 2011.

Colicchio, Tom. "Day 1: Our Afternoon at Anson Mills." *Food & Wine,* November 16, 2009. http://www.foodandwine.com/blogs/2009/11/16/Tom-Colicchio-Day-1-Our-Afternoon-at-Anson-Mills.

Columbia Convention and Visitors' Bureau. "5th Annual Viva La Vista." www.columbiacvb.com/includes/events/index.cfm?action=displayDetail&eventid=10643.

———. "Signature Events." www.columbiacvb.com/visitors/calendar-of-events/signature-events.

———. "3rd Annual Palmetto Tasty Tomato Festival." www.columbiacvb.com/includes/events/index.cfm?action=displayDetail&eventid=10642.

———. "2012 Irmo Okra Strut." www.columbiacvb.com/includes/events/index.cfm?action=displayDetail&eventid=10798.

DeWeerdt, Sarah. "Is Local Food Better?" Worldwatch Institiute. www.worldwatch.org/node/6064.

Discover Charleston. "South Carolina Gold Rush: The Resurgence of the State's Rice Culture." www.discovercharleston.com/cuisine/carolina-rice.htm.

Dunkle, David N. "Willie Nelson on the Farm Aid Cause: 'I Wish I Didn't Have to Do This.'" *Patriot-News*, September 21, 2012. www.pennlive.com/midstate/index.ssf/2012/09/willie_nelson_on_farm_aid_i_wi.html.

Eison, Joan Hardy. "Basil Garzia and Rosewood Market & Deli." *Healthy Living Columbia*, August 2012. www.healthylivingcolumbia.com/images/Rosewood.pdf.

Empire Records. Directed by Allan Moyle, 1995. Los Angeles, CA: Warner Brothers, 2003. DVD.

Farm Aid. "Farm Aid: Family Farmers, Good Food, a Better America." www.farmaid.org/site/c.qlI5IhNVJsE/b.2723609/k.C8F1/About_Us.htm.

———. "Farm Aid's 2012 Grants Focus on Growing Family Farm Agriculture and a New Generation of Farmers." December 12, 2012. www.farmaid.org/site/apps/nlnet/content2.aspx?c=qlI5IhNVJsE&b=2792875&ct=12532651&notoc=1.

———. "Helping Family Farmers Thrive." www.farmaid.org/site/c. qlI5IhNVJsE/b.2750749/k.89E0/Family_Farmers.htm.

Farm to School. "National Farm to School Network." www.farmtoschool.org.

Fowler, Gwen. "In the Kitchen with Chef Tim Peters." Discover South Carolina, April 27, 2012. www.discoversouthcarolina.com/Insider/ Food/Stories/8620.

Freshly Grown Farms. http://freshlygrownfarms.com.

Google Public Data. "Unemployment Rate—Not Seasonally Adjusted." January 20, 2013. https://www.google.com/ publicdata/explore?ds=z1ebjpgk2654c1_&met_y=unemployment_ rate&idim=city:PA450250&fdim_y=seasonality:U&dl=en&hl=en&q=cu rrent%20columbia%20sc%20unemployment%20rate.

Grace Communications Foundation. "Local & Regional Food Systems." www.gracelinks.org/254/local-regional-food-systems.

Harrelson, William L. *Handbook of South Carolina.* Columbia: South Carolina Department of Agriculture, 1973.

Harvest Dinners. www.harvestdinnersc.com/Welcome.html.

Hess, Karen. *The Carolina Rice Kitchen: The African Connection.* Columbia: University of South Carolina Press, 1992.

Infoplease. "South Carolina." *Columbia Encyclopedia.* 6th ed. www.infoplease. com/encyclopedia/us/south-carolina-history.html.

Irmo, South Carolina. "The Irmo Okra Strut." www.irmoinfo.com/okrastrut. html.

Jaws. Directed by Steven Spielberg, 1975. Martha's Vineyard, MA: Universal Pictures, 2000. DVD.

Jennings, Tenessa. "Motor Supply Co. Bistro Partners with Sustainable Farms for Summer Harvest Week." Lake Murray News, May 23, 2012.

http://lakemurray.wistv.com/news/arts-culture/52585-motor-supply-co-bistro-partners-sustainable-farms-summer-harvest-week.

———. "26th Annual Chili Cook Off in Five Points." Columbia News, October 31, 2012. http://columbia.wistv.com/news/community-spirit/54246-26th-annual-chili-cook-five-points.

Jones, Graham. "Live Aid 1985: A Day of Magic." CNN, July 6, 2005. http://edition.cnn.com/2005/SHOWBIZ/Music/07/01/liveaid.memories/index.html.

Kearns, Taylor. "Farmers Market to Open Next Saturday on Main Street." WISTV, October 12, 2012. http://www.wistv.com/story/19709492/farmers-market-to-open-next-saturday-on-main-street.

Kolar, Laura Richardson. "Historical Reflections of the Current Local Food and Agriculture Movement." *Essays in History.* Charlottesville: University of Virginia, July 2, 2011. www.essaysinhistory.com/content/historical-reflections-current-local-food-and-agriculture-movement.

Lexington County Peach Festival. http://lexingtoncountypeachfestival.com/index.html.

Local Harvest. "Community Supported Agriculture." www.localharvest.org/csa.

———. "D&J Farm." www.localharvest.org/d-j-farm-M9912.

———. "Doko Farm." www.localharvest.org/csa/M27508.

———. "Organic Food." www.localharvest.org/organic.jsp.

Maiser, Jennifer. "10 Reasons to Eat Local Food." "Life Begins at 30" blog, August 2005. www.lifebeginsat30.com/elc/2006/04/10_reasons_to_e.html.

McClam, Robbie. "City Roots: Your In-Town Sustainable Farm." Information sheet, January 3, 2013.

Moore, John Hammond. *Columbia and Richland County: A South Carolina Community, 1740–1990.* Columbia: University of South Carolina Press, 1993.

Motor Supply Company Bistro. "Bios." http://motorsupplycobistro.com/pages/bios.html.

Mr. Friendly's New Southern Café. http://mrfriendlys.com.

Opala, Joseph A. "South Carolina Rice Plantations." *The Gullah: Rice, Slavery, and the Sierra Leone—American Connection.* http://www.yale.edu/glc/gullah/02.htm.

OpenTable. Rosso Trattoria Italia. www.opentable.com/rosso-trattoria-italia.

Pig Business. www.pigbusiness.co.uk.

Pioreschi, Tom. "Farmers Market Huge Win for Main Street." *Free Times,* December 12, 2012. www.free-times.com/index.php?cat=1992912063981076&ShowArticle_ID=11011112121359450.

Rely Local. "Mr. Friendly's New Southern Cafe." www.relylocal.com/greater-columbia-south-carolina/business_listings/mr-friendlys-new-southern-cafe.

Rosewood Crawfish Festival. http://rosewoodcrawfishfest.com/index.html.

Rosewood Market. www.rosewoodmarket.com.

Rosso Trattoria Italia. http://rossocolumbia.com/index.html.

Round River Farms. www.roundriverfarms.com/overview.html.

Slow Food Columbia. www.slowfoodcolumbia.org.

Soda City Market. www.sodacitysc.com/default.html.

Solstice Kitchen & Wine Bar. www.solsticekitchen.com.

South Carolina Department of Agriculture. "SC Agribusiness." http://agriculture.sc.gov/content.aspx?MenuID=18.

South Carolina Department of Social Services. "Supplemental Nutrition Assistance Program (SNAP)." https://dss.sc.gov/content/customers/food/foodstamp/index.aspx.

South Carolina Farm to School. www.scfarmtoschool.com.

South Carolina Oyster Festival Facebook page. www.facebook.com/SCoyster.

South Carolina State Library. "A Brief History of South Carolina." www.statelibrary.sc.gov/a-brief-history-of-south-carolina.

Steffey, Hilde. "Oh Farm Bill, Where Art Thou?" Farm Aid Blog, January 29, 2013. http://blog.farmaid.org/2013/01/oh-farm-bill-where-art-thou.html.

Sustainable Midlands. www.sustainablemidlands.org.

Sustainable Seafood Initiative. "Sustainability Criteria." http://scaquarium.org/SSI/PDFS/sustainability_criteria.pdf.

———. "Sustainable Seafood Initiative Fact Sheet." http://scaquarium.org/SSI/PDFS/ssi_factsheet.pdf.

Terra. www.terrasc.com.

Trimarchi, Maria. "What Are Locavores?" TLC Cooking. http://recipes.howstuffworks.com/locavore1.htm.

United States Department of Agriculture. "Regional Food Hubs: Linking Producers to New Markets." www.ngfn.org/resources/ngfn-database/knowledge/RFHub%20Presentation_complete%20version_5.24.pdf.

University of South Carolina. "Sustainable Carolina." http://artsandsciences.sc.edu/greenquad.

————. "USC Alum Puts Columbia Farm on Slow Food Path," July 20, 2011. www.sc.edu/news/newsarticle.php?nid=1817#.UQv_k781k0M.

Vista Marketplace. http://vista.locallygrown.net.

Wil-Moore Farms. www.wil-moorefarms.com.

Worthy, Chris. "Adluh Flour Explores Social Media." *Columbia Business Monthly*, March 30, 2012. www.columbiabusinessmonthly.com/View-Article/ArticleID/1784/Adluh-Flour-Explores-Social-Media.aspx.

Index

U

University of South Carolina 13,
16, 17, 27, 28, 38, 50, 74, 93
Upstate 14, 15, 100

V

Vista Marketplace at Whaley 22,
62, 63, 100
Viva La Vista 30

W

Wil-Moore Farms 77, 78, 82, 90,
91
World War I 19, 28
World War II 21, 28, 37, 42, 43, 44

ABOUT THE AUTHOR

Laura Aboyan is a twentysomething food lover who is attempting to find the best places in Columbia, South Carolina, to eat, drink and be merry. She migrated south in August 2001 from suburban Philadelphia to attend college at the University of South Carolina and has not regretted that decision for a single second. She earned her bachelor's degree in public relations in 2005, and by August 2013, she will have earned her master's degree in higher education business administration, also from USC. She has been writing "The Hungry Lady" blog (the-hungry-lady.blogspot.com) since January 2011. When she isn't chronicling her food adventures, Laura spends most of her time cheering for the Gamecocks (as well as her hometown Phillies and Eagles), whining about grad school, going to see live music, reading anything she can get her hands on and trying to convince her friends to cook for her.

Visit us at
www.historypress.net
....
This title is also available as an e-book

www.ingramcontent.com/pod-product-compliance
Lightning Source LLC
Chambersburg PA
CBHW070342100426
42812CB00005B/1401